T0224983

SpringerBriefs in Applied Sciences and Technology

Computational Intelligence

Series Editor

Janusz Kacprzyk, Systems Research Institute, Polish Academy of Sciences, Warsaw, Poland

SpringerBriefs in Computational Intelligence are a series of slim high-quality publications encompassing the entire spectrum of Computational Intelligence. Featuring compact volumes of 50 to 125 pages (approximately 20,000–45,000 words), Briefs are shorter than a conventional book but longer than a journal article. Thus Briefs serve as timely, concise tools for students, researchers, and professionals.

More information about this subseries at http://www.springer.com/series/10618

Manasvi Aggarwal · M. N. Murty

Machine Learning in Social Networks

Embedding Nodes, Edges, Communities, and Graphs

 Springer

Manasvi Aggarwal
Department of Computer
Science and Automation
Indian Institute of Science
Bengaluru, Karnataka, India

M. N. Murty
Department of Computer
Science and Automation
Indian Institute of Science
Bengaluru, Karnataka, India

ISSN 2191-530X   ISSN 2191-5318   (electronic)
SpringerBriefs in Applied Sciences and Technology
ISSN 2625-3704   ISSN 2625-3712   (electronic)
SpringerBriefs in Computational Intelligence
ISBN 978-981-33-4021-3   ISBN 978-981-33-4022-0   (eBook)
https://doi.org/10.1007/978-981-33-4022-0

This Springer imprint is published by the registered company Springer Nature Singapore Pte Ltd.
The registered company address is: 152 Beach Road, #21-01/04 Gateway East, Singapore 189721, Singapore

Preface

Overview

Network analysis has gained a lot of prominence over the past decade. This is because of a better understanding and control over *learning representations* of various entities like nodes, edges, subgraphs, cliques, and graphs that represent important components of a network.

- *Representation* itself is an important task in a variety of tasks. Typically, networks are represented as graphs. In turn, graphs are represented using *adjacency matrices*.
- A related major breakthrough in the recent past is based on *representing words as vectors*. This was extended to representing nodes in a graph by using *random walks* over nodes in the graph as sentences. This view facilitated representing nodes as vectors.
- Another popular technique that can be viewed as an equivalent to random walk-based schemes uses *matrix factorization*.
- One more important area is *deep learning* based on deep neural networks that have contributed significantly to the state of the art in the topics of interest to this book.

This book deals with *network embedding*. There are different embedding schemes that will be discussed in the book.

- Networks, especially *social networks* and their generic properties that help in *network representation learning*, will be examined.
- Backgrounds required to deal with *social network analysis* in the form of neural networks and deep learning are also important, and they will be discussed.
- *Node embedding* schemes based on random walks, matrix factorization, and deep learning will be covered along with the state-of-the-art developments.
- There are several applications that require embedding of entities beyond nodes. Applications in the areas of social, health, finance, education, and transportation networks may require embedding of cliques, subgraphs, and graphs. *Embedding*

graphs is one of the active research areas now. Some of the state-of-the-art algorithms in this context will be covered in this book.
- Typically, the embeddings obtained are evaluated using several downstream machine learning (*ML*) tasks including *classification, community detection, visualization, and link prediction in social and information networks.*

Audience

This book is intended for senior undergraduate and graduate students and researchers working in social and complex networks. We present material in this book so that it is accessible to a wide variety of readers with some basic exposure to undergraduate-level mathematics. The presentation is intentionally made simpler for the comfort of the reader.

Organization

This book is organized as follows:

Chapter 1 deals with a generic introduction to social network embedding. Chapter 2 deals with social networks, their representation using graphs, and various embedding schemes. It also deals with some important topics like the downstream *ML* tasks including classification, clustering, visualization, and link prediction. Chapter 3 provides a coverage on neural networks and popular deep learning tools including convolutional neural networks (*CNNs*), recurrent neural networks (*RNNs*), and autoencoders. Node representation forms the subject matter of Chap. 4. Embedding graphs is examined in Chap. 5. Finally, it is concluded in Chap. 6.

Bengaluru, India   Manasvi Aggarwal
                                                                            M. N. Murty

Contents

Acronyms

AI	Artificial Intelligence
CNN	Convolutional Neural Network
DGI	Deep Graph Infomax
GAT	Graph Attention Network
GCN	Graph Convolution Network
GIN	Graph Information Network
GNN	Graph Neural Network
JC	Jaccard Coefficient
LP	Link Prediction
LSTM	Long Short Term Memory
ML	Machine Learning
MLP	Multi Layer Perceptron
NMF	Nonnegative Matrix Factorization
NRL	Network Representation Learning
RNN	Recurrent Neural Network
SIN	Social and Information Network
SVD	Singular Value Decomposition
TADW	Text Attributed Deep Walk
WL	Weisfeiler-Lehman

Chapter 1
Introduction

1.1 Introduction

Networks have the capacity to represent and solve many real world problems and consequently, their analysis is gaining prominence. In practice, there are several applications in which the underlying network is *explicit*. Examples include a friendship network, a citation network, and the world wide web. It is also possible to view several other applications using an *implicit* network. For example, in an artificial intelligence application, the data might be given in the form of a set of vectors, or sequences. Even in such applications, it may make sense to *derive* the network structure *implicitly present* in the seemingly isolated set of data points.

Networks allow us to exploit the domain knowledge. Analysing networks will help in detecting both the *latent content and structural dependencies* between entities.

- Networks can effortlessly handle the relational complex systems underlying various applications in education, biological, citations or societal domains.
- Network structure contains properties of the corresponding data such as connections encode the similarity between network entities. In a social network if two users are connected directly then those two users are friends of each other.
- To analyse the hidden network properties and the data that the network is representing, various machine learning and AI tools are used.
- But all machine learning (ML) algorithms cannot process the networks in their raw form. They need vectors of numbers as input.
- Also, network data is high dimensional and therefore, analysis becomes challenging.
- To solve all these and related issues embedding tools are used. Network Representation Learning (NRL) algorithms are developed to convert the raw structure to a real-valued low-dimensional vector called its embedding.

M. Aggarwal and M. N. Murty, *Machine Learning in Social Networks*,
SpringerBriefs in Computational Intelligence,
https://doi.org/10.1007/978-981-33-4022-0_1

– These NRL techniques capture different properties of the network and these vectors are then exploited by various ML tasks which further help us both in analysing the network and validating the effectiveness of the vectors.
– Conventional ways represent networks in high dimensional spaces. Further, they cannot embody relations between nodes that are not adjacent.
– Embeddings can embody different properties of the network and information. Thereby they show high performance on downstream ML tasks.

• We use an example to illustrate the idea of embedding.

– Consider a network represented as a graph in Fig. 1.1. There are six nodes in the network and are represented using six vertices labeled $a, b, c, d, e,$ and f.
– Such a graph may be represented as an *adjacency matrix* having one row for each vertex and a column for each of the vertices.
– So, the adjacency matrix A is a 6×6 matrix. The value of $A_{i,j}$, the element in the ith row and jth column is 1 if there is an edge between nodes i and j where $i, j = a, b, \ldots, f$.
– The entire adjacency matrix is given by

$$A = \begin{bmatrix} 0 & 1 & 1 & 0 & 0 & 0 \\ 1 & 0 & 1 & 0 & 0 & 0 \\ 1 & 1 & 0 & 1 & 0 & 0 \\ 0 & 0 & 1 & 0 & 1 & 0 \\ 0 & 0 & 0 & 1 & 0 & 1 \\ 0 & 0 & 0 & 1 & 1 & 0 \end{bmatrix}$$

– In the matrix rows correspond respectively to $a, b, c, d, e,$ and f. Similarly columns are ordered based on a to f in the order.
– Each row may be viewed as a binary vector representing the edges between the corresponding node and all the vertices. For example, the first row corresponds to the edges between a and all the vertices.
– There are no self loops here. So, all the diagonal entries in the matrix have a value 0 (zero).
– We can embed these six dimensional vectors using the following observations.

Fig. 1.1 Adjacency list of the network in figure

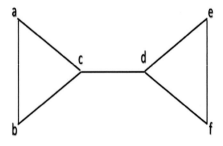

- Observe that in Fig. 1.1, nodes a, b, and c belong to one connected component and nodes d, e, and f belong to the other component.
- We can record these assignments by using the following component assignment matrix

$$CAM = \begin{bmatrix} 1 & 0 \\ 1 & 0 \\ 1 & 0 \\ 0 & 1 \\ 0 & 1 \\ 0 & 1 \end{bmatrix}$$

Here the first column corresponds to the first component and the second column corresponds to the second component.

- Each point is a six-dimensional binary vector and it belongs to one of the connected components in this simple case.
- Here the matrix CAM may be viewed as storing the two dimensional representations of each of the points in a component.
- This matrix CAM may be viewed as an **embedding matrix** providing a two-dimensional embedding of the six vertices.
- Note that similar vertices share the same embedding in this example. Nodes a, b, and c are similar as they belong to the same connected component. Further, d, e, and f belong to the same connected component and are similar.
- Points a, b, c are represented using the vector $(1, 0)$ and the remaining three vertices are mapped to the vector $(0, 1)$.
- In this example, the six dimensional data points are mapped to a two-dimensional space. In general if there are N nodes in a network, then the adjacency matrix will be an $N \times N$ matrix.
- By using the centroid or mean of the three six-dimensional vectors in each component as the representative of the three points, we get the representative matrix as

$$RM = \begin{bmatrix} \frac{2}{3} & \frac{2}{3} & \frac{2}{3} & \frac{1}{3} & 0 & 0 \\ 0 & 0 & \frac{1}{3} & \frac{2}{3} & \frac{2}{3} & \frac{1}{3} \end{bmatrix}$$

Note that the first row of RM is the centroid of the first group of 3 points and the second row is the centroid of the second group of 3 points.

- It is possible to view the product of CAM and RM as an approximation to A. That is

$$A \approx (CAM)(RM)$$

- In the example

$$
\begin{bmatrix} 0\ 1\ 1\ 0\ 0\ 0 \\ 1\ 0\ 1\ 0\ 0\ 0 \\ 1\ 1\ 0\ 1\ 0\ 0 \\ 0\ 0\ 1\ 0\ 1\ 0 \\ 0\ 0\ 0\ 1\ 0\ 1 \\ 0\ 0\ 0\ 1\ 1\ 0 \end{bmatrix} \approx \begin{bmatrix} 1\ 0 \\ 1\ 0 \\ 1\ 0 \\ 0\ 1 \\ 0\ 1 \\ 0\ 1 \end{bmatrix} \begin{bmatrix} \frac{2}{3}\ \frac{2}{3}\ \frac{2}{3}\ \frac{1}{3}\ 0\ 0 \\ 0\ 0\ \frac{1}{3}\ \frac{2}{3}\ \frac{2}{3}\ \frac{1}{3} \end{bmatrix}
$$

- In a general case of an $N \times N$ matrix, if the N dimensional points are assigned to K groups, then CAM is an $N \times K$ matrix and the corresponding RM is a $K \times N$ matrix.
- For example, such a representation can be obtained by using some clustering or grouping algorithm like the K-means clustering algorithm on the rows of matrix A.

- In general there could be many other ways of embedding nodes or edges in a graph.
- This book deals with network embeddings from many aspects such as embedding nodes, edges as well as complete graphs.
- Also, division of all the techniques based on whether the learning is *supervised and unsupervised*.
- In an unsupervised setting, we do not need the *class labels* of each of the points for learning. For example, in the example graph shown in Fig. 1.1, we have not explicitly used any classes and their labels.
- We have grouped the points based on their being members of a component or not.
- It is possible to convert the above example to lead to a supervised learning problem by saying, for example, that points a, *and* c are in $class - 1$ and d, *and* f are in $class - 2$.
- In such a case, we can still talk of a class-assignment matrix $CLAM$ as a 6×2 matrix given below:

$$
CLAM = \begin{bmatrix} 1\ 0 \\ ?\ ? \\ 1\ 0 \\ 0\ 1 \\ ?\ ? \\ 0\ 1 \end{bmatrix}
$$

- Now the first column of $CLAM$ corresponds to $Class - 1$ and the second column corresponds to $Class - 2$.
- Here also the rows correspond to the six points in the order a to f.
- Note that there are question marks in the rows corresponding to the assignment of points b and e in $CLAM$.
- This is because we do not know the class labels of b and e.
- One way to classify these points is to use a simple classification rule called the nearest neighbor (NN) rule.
- NN rule says that assign a node to the class of its nearest neighbor.

- Further, for a node v_i let $NN(v_i)$ be a node v_j if v_j, $(i \neq j)$ has the shortest path length from v_i to any node.
- Note that $NN(b)$ is either a or c as both are at a distance of 1 unit from b as they are directly connected to b. In this simple case, both a and c belong to $Class - 1$; so b is also assigned to the same class.
- Similarly, node e is assigned to $Class - 2$ because $NN(e)$ is either d or f. Again both are from $Class - 2$. So, e is assigned to $Class - 2$.
- In this simple supervised example, the matrix $CLAM$ may be updated to indicate the labels assigned to nodes b and e using the labels of their NNs.
- The updated $CLAM$ is

$$CLAM_u = \begin{bmatrix} 1 & 0 \\ 1 & 0 \\ 1 & 0 \\ 0 & 1 \\ 0 & 1 \\ 0 & 1 \end{bmatrix}$$

- The above discussion based on a simple example is considered only to illustrate various notions.
- More generic application scenarios may require more complex schemes for generating embeddings.
- Such state-of-the-art approaches are discussed later in the book and also a comparison of these schemes will be considered.
- In addition, details of evaluation metrics for each category are covered.

1.2 Notation

This section summarizes the notation used in the context of graphs throughout the book. The details are provided in Table 1.1. Notation specific to a topic are excluded to avoid confusion. Also, the related terminology is discussed when it is needed.

1.3 Contents Covered in This Book

- Chapter 2 deals with an introduction to *Network Representation Learning (NRL)* followed by a discussion on the datasets and some *machine learning downstream tasks* that are being used to evaluate the effectiveness of embedding learning approaches. Also, matrix factorization approaches and word2vec model will be presented.
- Chapter 3 covers *basics of Deep Learning* that aid in network representation learning and analysis tasks that will be covered later in this book. Multi-Layer neural

Table 1.1 Summary of graph notation used in the book

Notation	Explanations
$G = (V, E)$	Graph
V	Set of nodes of the graph
E	Set of edges of the graph
N	Number of nodes in the graph
A	Adjacency matrix of the graph
v_i	ith node of graph
e_{ij}	An edge between nodes v_i and v_j
x_i	Attribute associated with ith node
X	Set of node attributes
l_i	Label associated with the ith node
L	Set of node labels
$\mathcal{G} = \{G_1, \ldots, G_M\}$	Set of graphs
L_g	Set of graph labels
$x^i_j \in \mathbb{R}^D$	Attribute vector for the jth node in the ith graph

networks, convolutional neural networks, recurrent neural networks and autoencoders are covered in this chapter.

- Chapter 4 presents *node embeddings*. These algorithms are categorized based on the techniques they use. For example, Matrix Factorization approaches and Deep Learning based approaches that exploit Graph Neural Networks.
- Chapter 5 details the *Graph Embedding* approaches. These approaches are categorized based on whether graph embeddings are generated by first learning node embeddings or directly without using the node embeddings. Subsequently, graph pooling techniques are examined.

Chapter 2
Representations of Networks

2.1 Introduction

Networks are becoming ubiquitous as they can represent many real-world relational data, for instance, information networks, molecular structures, telecommunication networks, and protein-protein interaction networks.

A *Network* is a collection of entities and feasible connections between them. A Network is most commonly represented using a *Graph*. Network analysis is carried out by analysing the underlying graph. Even in non-network applications where the data are not explicitly linked, it is possible and helpful to represent the data in the form of a network/graph. For example, in probabilistic graphical models, the data is visualized as a graph.

A *graph* is a non-euclidean data structure which is represented by $G = (V, E)$, where V is the set of nodes and E is the set of edges. The *nodes* represent the network entities and *edges* represent the connections between the entities. An edge $e_{ij} \in E$ between two nodes v_i and v_j is represented by a pair of the nodes (v_i, v_j). For example, in a social network, each node v_i represents a user, and an edge (v_i, v_j) represents whether user v_i is friends with user v_j or not. In protein-protein interactions network, nodes represent proteins and edges represent interactions between these proteins. Also, some real-world graphs have an associated set of attributes, where each node $v_i \in V$ is associated with an attribute vector $x_i \in \mathbb{R}^D$, D is the dimension of each attribute vector. Further, graphs can also have an associated node labels set L, where $y_i \in L$ is the label of node v_i, which specifies the class of the node.

Analysis of these networks provides advantages in many fields such as *recommendation* (recommending friends in a social network), biological field (deducing connections between proteins for treating new diseases), *community detection* (grouping users of a social network according to their interests), etc. by leveraging the latent information of networks. Hence, network analysis is gaining prominence. But the high dimensional, irregular graph data imposes challenges for machine learning tasks which led to the development of many representation learning techniques.

© The Author(s), under exclusive license to Springer Nature Singapore Pte Ltd. 2021
M. Aggarwal and M. N. Murty, *Machine Learning in Social Networks*,
SpringerBriefs in Computational Intelligence,
https://doi.org/10.1007/978-981-33-4022-0_2

2.2 Networks Represented as Graphs

Any network can be easily represented as a *Graph*, defined in Sect. 2.1, which facilitates modeling data items and relations among them.

Many variants of *Graphs* are possible including:

- *Heterogeneous Graphs*: The nodes or/and edges of such graphs can be of various types and each type must be handled differently.

- *Homogeneous Graphs*: Contrary to heterogeneous graphs, nodes and edges are instances of a single type.

- *Directed and Undirected Graphs*: Directed graphs have ordered pairs of vertices, and each edge has a starting point (head) and an ending point (tail), and information flows from head to tail. In contrast, an edge in an undirected graph can be traversed in both directions representing a symmetric relation. Further, an undirected edge can be replaced with two directed edges. In a directed graph, an edge between nodes v_i and v_j is represented by (v_i, v_j), whereas in the undirected graph, it can be written either way i.e., (v_i, v_j) or (v_j, v_i).

- *Dynamic Graphs*: Some real-world networks might evolve over time. For example, in social networks, new users can be added, or new interactions might occur between existing users. This leads to the addition or removal of nodes or edges, respectively, and hence these graphs are called *dynamic*. On the contrary, graphs which do not change over time are known as *static graphs*.

- *Knowledge Graphs*: A Knowledge Graph is a directed, multi-relational graph where an edge is represented in (head entity (h), relation (r), tail entity (t)) form, which means that h is related to t through r. For instance, (Star Trek, Genre, Science Fiction).

- *Hypergraphs*: They are the generalization of undirected graphs in which edges are over subsets of two or more vertices. Formally, a hyper-graph H is a pair $H = (X, E)$ where X is a set of elements called nodes or vertices, and E is a set of non-empty subsets of X called hyper-edges.

All these variants might contain useful auxiliary information such as vertex attributes and/or vertex labels, in addition to the connectivity/structural information.

A toy example of an undirected and homogeneous graph with 7 nodes and 9 edges is depicted in Fig. 2.1. To understand some important properties of a graph, let us observe the same.

1. This example illustrates an undirected and static homogeneous graph. Therefore, edges can be represented in both directions. For example, an edge between *a* and

Fig. 2.1 A Toy network:
Circles $(a, b, \ldots, g)$ are the
nodes of the graph while
black lines denote the edges

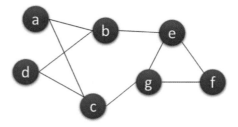

b can be denoted as (a, b) or (b, a). In a friendship network, if u_1 and u_2 *are connected* then stating that u_1 *is a friend of* u_2 *is the same as* u_2 *is a friend of* u_1.

2. *Neighbours* are those nodes that are connected by an edge. For example, node a and node b are connected by an edge and hence are called neighbors (also called adjacent nodes) of each other.

3. *Common Neighbor:* Further observe that both nodes a and b have an edge to node c, that is why c is their *common neighbour*.

4. *Degree* of a node is the total number of incident edges on that node in an *undirected graph*. The degree of node a is 2. Further, the sum of the degrees is equal to 2 times the number of edges:

$$\sum_{i=1}^{|V|} deg(v_i) = 2|E| \tag{2.1}$$

where $|V|$ is the total number of nodes in the graph.

5. In directed graphs, the degree of any node v is the sum of indegree and outdegree of v, where *indegree* is the number of incoming edges on v and *outdegree* is the number of edges leaving v. Also, the total indegree of the graph is equal to the total outdegree of the graph.

A graph can be represented in numerous ways. All variants, as described in Sect. 2.2, need different representation schemes. In this book, our focus is on undirected/directed, static homogeneous graphs. Next, we discuss the two most commonly used methods to represent graphs on the machine.

2.3 Data Structures to Represent Graphs

2.3.1 Matrix Representation

Adjacency Matrix (A): It is a square matrix where number of rows and number of columns are the same as the number of nodes in the graph, i.e. the dimension of A is $N \times N$ where N is the number of nodes. Each (i, j)th entry in the matrix indicates

Table 2.1 Adjacency matrix for the graph in Fig. 2.1

From/To	a	b	c	d	e	f	g
a	0	1	1	0	0	0	0
b	1	0	0	1	1	0	0
c	1	0	0	1	0	0	1
d	0	1	1	0	0	0	0
e	0	1	0	0	0	1	1
f	0	0	0	0	1	0	1
g	0	0	1	0	1	1	0

presence or absence of an edge between nodes v_i and v_j. If $A_{i,j}$ is 1 then nodes v_i and v_j are connected by an edge otherwise 0. If the graph is weighted then each entry of the matrix will store the weight of the corresponding edge.

Table 2.1 describes the matrix A for graph in Fig. 2.1. Some important points to observe from A are:

- Each node v_i of a graph is given an index i and ith row of matrix A corresponds to v_i.
- The graph in Fig. 2.1 has 7 nodes; therefore, dimension of matrix A is 7×7.
- As the graph is undirected, A is a symmetric matrix. For a directed graph, the adjacency matrix may not be symmetric.
- *Degree* of a node for an undirected graph can be computed by taking the sum of entries in either the respective row or the column of matrix A. Degree of node a is 2 as

$$\sum_{i=0}^{6} A_{i,0} = \sum_{i=0}^{6} A_{0,i} = 2 \qquad (2.2)$$

- To store matrix A, $O(N^2)$ space is required, where N is the number of nodes.
- Observe that all diagonal entries are 0. If a graph has a self loop on a vertex v_i, then A_{ii} will be equal to 1, and each self loop contributes 2 to its degree (incoming and outgoing is the same vertex).
- Also note that matrix A of a network abstracts paths of length 1, $A \times A$ gives paths of length 2 and so on.

2.3.2   Adjacency List

Another way to represent a graph is by using its Adjacency List (adjlist). It is an array of linked lists where the ith list, adjlist[i], stores references to every neighbour of node v_i (Fig. 2.2).

Fig. 2.2 Adjacency list of
the network in Fig. 2.1

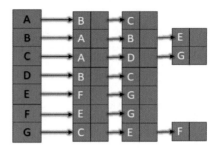

- Each node v_i of a graph is given an index i and the respective linked list stores its adjacent nodes. As can be observed from Fig. 2.1, node a (given 0th index) has nodes b and c as its neighbors and hence adjlist[0] has pointer to the linked list (an array can also be used instead of linked list) which stores references to nodes b and c. Also, the number of linked lists is equal to the number of nodes.
- Another interesting point is that if $(v_i, v_j) \in E$ in an undirected graph, then the list of v_i will have an entry for v_j and vice versa, whereas if it is an edge in a directed graph, then only v_i will have an entry for v_j in its list.
- For weighted graphs, each node entry will have two items, one for node index and the other one for the respective edge weight.
- A further observation is that the length of a list tells the degree of the respective node in undirected graphs.

2.4 Network Embeddings

Because of the prominence of networks in many real-world problems, network analysis is gaining importance in many disciplines. Networks contain a pool of information needed by all end stream tasks.

Due to this, analysis schemes depend substantially on network representations (also known as *encodings or embeddings*). Furthermore, structural information, which may be intractable, has to be inscribed in the low dimensional vector space. For example, classification of a node requires the global position and the local neighborhood structure of the node; link prediction task needs to capture the similarity between two or more nodes. But in some application domains, the network structure may not be apparent.

Therefore, we need an efficient and optimal way to convert the raw non-euclidean high dimensional network data to a vector such that implicit or explicit relations/properties present in the network are protected in the vector space. For example, distance between nodes in a network should be preserved in the embedding space. *Embedding learning* is also called *representation learning*.

Network Representation Learning: Given a network represented by a graph $G=(V,E)$ and some side information related to G such as a set L of discrete labels to label the nodes and a matrix of vertex attributes X. The aim is to learn a mapping, $g_v : V \mapsto Z$, that maps each node v_i of G to a vector z_i of dimension d known as embedding.

- Conventional ways of using adjacency vector for representing nodes are not appropriate as large scale data makes computation intractable in high dimensional spaces. Also, they capture only first-order dependencies and are incapable of including higher-order relations of a network.
- However, network representation learning methods learn explanatory embedding vectors because of which many machine learning analysis tasks such as recommendation, link prediction, node and graph classification, community detection, and visualization can be efficiently tackled using these embeddings.
- Important properties that embeddings must encode are explained below.

  - As discussed earlier, some real-world graphs are augmented with node attributes that help network representation learning techniques to learn more discriminative embeddings as these approaches also capture the attribute level similarity between nodes.
  - Other common properties of interest are the relationships of various order of proximities between nodes.

    · The first-order proximity captures edges of a graph. Thus, if v_i, v_j are connected by an edge then the first-order proximity is 1 between v_i and v_j. And it is 0 for non-adjacent pairs of nodes. This information can easily be gathered from the adjacency matrix of the graph.
    · Two-hop relations are described by the *second-order proximity* which is measured by the number of 2-hop paths between v_i and v_j. These paths between a pair of nodes can be calculated using the second-order adjacency matrix A^2.
    · Similarly, *higher-order proximity* captures node pairs with q-hop paths where q is greater than or equal to 3. This is determined by q-step transition probability i.e., the probability of landing on node v_j at qth step of a random walk starting from node v_i. As the order of proximity increases, the ability to capture global structure also increases.

  - Moreover, many nodes of a graph share the common responsibility such as nodes playing the role of a hub node. This property is also known as the *structural equivalence* property.
  - Another important principle is homophily, i.e. the nodes of a graph form a community structure. All vertices in a single community share some common interest or property. For example, in citation networks, all papers in a single community are on the same or a similar topic.
  - An acceptable representation should exhibit a proper balance of both of these equivalences i.e., embeddings of nodes from the same community should be

similar to each other and the nodes with similar structural roles should also be
embedded closer to each other.

2.5 Experimental Datasets and Metrics

Many Benchmark datasets are being used to evaluate the quality of the network
embeddings or performance of algorithms on various downstream tasks.

- Some datasets contain information of a single graph, e.g. adjacency matrix, node
  attributes, node labels, etc. They are used for node or edge level analysis, such as
  node classification, and link prediction.
- Some datasets are a collection of graphs and are used to evaluate the efficiency
  of the graph level embeddings on graph level tasks, such as graph classification,
  graph clustering, etc.

This section briefs some of the most commonly used datasets and the evaluation
metrics for various downstream tasks.

2.5.1 Evaluation Datasets

Many datasets are publicly available for measuring the effectiveness of algorithms
and comparing them with the state-of-the-art approaches depending on the performance on various downstream tasks. This section discusses some of the most frequently used datasets.

1. *Citation networks*:
   These datasets are collections of scientific, academic publications with nodes
   denoting authors or papers and edges representing author-coauthor or paper-paper
   citation relations. Examples of citation networks include Cora, Citeseer, PubMed
   datasets. Below is a brief description of these datasets (Table 2.2).

   - Cora, Citeseer, PubMed are unweighted and directed networks and have node
     attributes that denote the contents of the papers or authors.
   - The number of distinct class labels varies across the datasets. Cora has seven
     class labels, Citeseer has six distinct node labels, while PubMed has three
     distinct node labels.

2. *Biological networks*:

   - *Single graph Dataset:* A PPI (Proteins-Proteins Interaction) dataset is from
     this category whose nodes represent proteins and edges represent the existing
     physical interactions. It has 40 different node labels, with each class denoting
     some biological state.

- *Collection of networks (graphs):* Each dataset has information for multiple networks (graphs). Some frequently used datasets are MUTAG, PROTEINS, NCI1/NCI109, ENZYMES, etc. All these datasets have an adjacency matrix for each graph, graph identifiers for all the nodes, and graph labels for all the graphs. Further, this information can be supplemented by node attributes, edge labels, and graph attributes (Table 2.3).

  - *MUTAG:* It is a collection of chemical compounds having different mutagenic behavior on a bacterium. It contains information of 188 graphs categorized into two classes.
  - *PROTEINS:* This dataset is a collection of proteins, where graphs are secondary structure elements. There is an edge between two nodes if the nodes are in any sequence of amino acids. The number of graphs is 1113, and the number of graph labels is two.
  - *ENZYMES:* Each graph in the set is a protein tertiary structure. It is a collection of 600 graphs, with each graph categorized into one of the two classes.

  Other bioinformatic datasets are PTC, FRANKENSTEIN, etc. The details of these datasets can be found at (https://bit.ly/39T079X). Table 2.3 contains a high-level summary of these datasets.

3. *Social Networks:*

   - *Single graph Dataset:* Some commonly used datasets are YouTube, Flickr, BlogCatalog, where nodes correspond to the users of that social website, and edges describe the relations between users of the website. For instance, nodes of the BlogCatalog network represent bloggers. Important statistics are provided in Table 2.4.
   - *Collection of networks (graphs):* Datasets such as IMDB-BINARY, IMDB-MULTI, COLLAB, REDDIT (Binary and Multi), etc. are the most commonly used graph level social network datasets. The details of these datasets can be found at (https://bit.ly/39T079X). Refer to Table 2.5 for a high-level description of these datasets.

4. *Collaboration Networks*: Arxiv is a collaboration network formed from the ArXiv website, and edges represent the co-author relations. Papers are only from a single field, and therefore the corresponding node labels are absent. Missing node label information makes this dataset suitable for link prediction.

There are many other types of datasets, e.g., Language Networks (Wikipedia), Communication Networks (Enron Email Network), etc.

2.5.2 Evaluation Metrics

As discussed, the embeddings are used for various downstream tasks (detailed discussion in Sect. 2.6). The performance on these tasks throws light on the algorithm's

Table 2.2 Citation networks for node level experiments

Dataset	#Nodes	#Labels	Attributes
Cora	2,708	7	Yes
Citeseer	19,717	6	Yes
PubMed	3,312	3	Yes

Table 2.3 Different biological datasets used in graph level experiments

Dataset	#Graphs	#Max Nodes	Avg.#Nodes	Avg. #Edges	#Labels	Attributes
MUTAG	188	28	17.93	19.79	2	NO
PTC	344	64	14.29	14.69	2	NO
ENZYMES	600	125	32.63	62.14	6	Yes
PROTEINS	1113	620	39.06	72.82	2	Yes
DD	1178	5748	284.32	715.66	2	NO
NCI1	4110	111	29.87	32.30	2	NO
NCI109	4127	111	29.68	32.13	2	NO
FRANKENSTEIN	4337	214	16.90	17.88	2	YES

Table 2.4 Social networks for node level experiments

Dataset	#Nodes	#Labels	Attributes
BlogCataloga	10,312	39	NO
Flickrb	80,513	195	NO
YouTubeb	1,138,499	47	NO

Table 2.5 Social graph level datasets

Dataset	#Graphs	#Max Nodes	Avg. #Nodes	Avg. #Edges	#Labels	Attributes
IMDB-BINARY	1000	136	19.77	96.53	2	NO
IMDB-MULTI	1500	89	13.00	65.94	3	NO
COLLAB	5000	492	74.49	2457.78	3	NO
REDDIT(BINARY)	2000	3782	429.63	497.75	2	NO
REDDIT(MULTI)-12K	11929	3782	391.41	456.89	11	NO

efficacy and helps in comparing different algorithms. This section details the metrics which are popularly used to evaluate the performance of the embeddings through algorithms on different downstream tasks.

- *Classification Accuracy* is the simplest metric which tells how many correct predictions a model makes i.e., $\frac{\#Correct\ Predictions}{\#Total\ Samples}$. This measure may not be useful when the data has *class imbalance*, that is most of the training patterns are from one class and the other class(es) have a very small number of training patterns.

- *F1 score* is the weighted average of precision (P) and recall (R) and its value lies between 0 and 1, with 1 being the highest score i.e., a model needs to maximize the F1-score.

$$F1_{micro} = 2 * (P * R)/(R + R) \tag{2.3}$$

This is also known as *F1-micro score*. Another score is *F1-macro*, which is defined as

$$F1_{macro} = \frac{\sum_{y \in L} F1(y)}{|L|} \tag{2.4}$$

Here $y \in L$ is the node label, $F1(y)$ is the $F1$-score for label y and $|L|$ is the number of distinct node labels.

- *Precision and Recall:*

  · *Precision* calculates how precise the model is or how many are actually positive (true positive) among all the predicted positives. This metric is useful when misclassifying negative sample costs more. Precision is defined as:

  $$Precision(P) = \frac{\#True\ Positives}{\#True\ Positives + \#False\ Positives} \tag{2.5}$$

  · *Recall* calculates how many samples from the positive class the model can predict correctly. This is used when misclassifying a positive class sample costs more to the user. Recall is described as:

  $$Recall(R) = \frac{\#True\ Positives}{\#True\ Positives + \#False\ Negatives} \tag{2.6}$$

  Here, true positives are those inputs that belong to the positive class and are classified correctly. False positives are those samples which belong to the negative class but are misclassified as positives. Similarly, true negatives belong to the negative class and are correctly classified, while false negatives belong to the positive class and are misclassified.

- *Precision at k (Pr@k)* : Instead of evaluating the model with respect to all the samples, it calculates the correct number of predictions in only the top k predicted edges:

$$Pr@k(i) = \frac{|\{v_j | v_i, v_j \in V, (v_i, v_j) \in E, rank(v_j) \leq k\}|}{k} \tag{2.7}$$

Here V is the set of nodes, E is the set of edges, $v_i, v_j \in V$ are the nodes, and rank(.) is the rank of the node. This metric is used to measure the efficiency of algorithms on the link prediction task.

- *NMI (Normalized Mutual Information)*: It is a normalized (i.e., between 0 and 1) mutual information (MI) score with 0 denoting no MI and 1 being the perfect correlation. This score is permutation invariant; thus, the score will remain the same for any permutation of the cluster labels. NMI is defined as:

$$NMI(L, W) = \frac{I(L; W)}{\frac{[H(L)+H(W)]}{2}} \qquad (2.8)$$

Here L is the set of actual labels (ground truth clusters based on the node labels), W is the set of predicted labels based on the predicted clusters, $H(.)$ is the entropy function, and $I(., .)$ is mutual information.

- *Entropy of class labels (H(Y))* talks about the uncertainty and is calculated using the following equation:

$$H(L) = -\sum_{i \in L} P(L = i) \times \log P(L = i) \qquad (2.9)$$

- Entropy of cluster labels (H(W)) is also calculated similarly:

$$H(W) = -\sum_{i \in W} P(W = i) \times \log P(W = i) \qquad (2.10)$$

Here P(W=i) is given by $\frac{\#Samples\ in\ cluster\ i}{Total\ Number\ of\ Samples}$.

- The next equation describes how to calculate the MI between the class labels and the cluster labels, which denotes the entropy reduction of class labels when cluster labels are given.

$$I(L; W) = H(L) - H(L|W) \qquad (2.11)$$

Here H(L|W) is the conditional entropy of class labels, and for each cluster i, it is calculated using the following equation:

$$H(L|W = i) = -P(W = i) \sum_{l \in L} P(L = l|W = i) \log(P(L = l|W = i))$$
$$(2.12)$$

Here P(L=l|W=i) is the probability of getting sample with label l in cluster i.

Putting all these values together, we can get the final NMI score using Eq. 2.8.

- *Purity* is a simple clustering measure that determines how many samples are correctly placed after clustering. All samples in a cluster are given a single label according to the most frequent label in that cluster. The fraction of correctly labeled samples to the total samples is known as the Purity.

Fig. 2.3 A Toy Network that depicts the results of Machine Learning tasks. (1) All nodes are colored according to the node labels such as all blue color nodes have same node label. All grey colored nodes are unlabeled. (2) Dotted circles denote a cluster. All nodes within a dotted circle belong to the same cluster or community. (3) Dotted lines are the predicted links

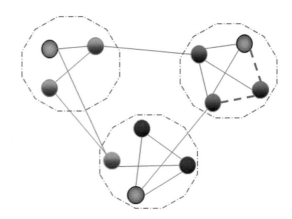

2.6 Machine Learning Downstream Tasks

One of the important factors behind learning low-dimensional network embeddings is that the same embedding can be exploited to deal with a host of *machine learning (ML)* tasks. These downstream *ML* tasks are performed on top of the trained embeddings or in an end-to-end fashion while learning embeddings.

- These *ML* tasks include *classification, clustering, link prediction, visualisation, and reconstruction.*
- Any off-the-shelf machine learning algorithm can perform the tasks mentioned above using the learned embeddings as input features.
- The performance on these machine learning tasks is further used for a fair comparison of approaches and as a measure to evaluate the quality of the learned embeddings.

In this section, we will explain these tasks with the help of Fig. 2.3 and provide their respective evaluation measures.

2.6.1 Classification

This is the most commonly used downstream task to justify the effectiveness of the trained embeddings.

- In practice, some networks are augmented with node labels or edge labels or graph labels categorizing these network entities into distinct categories. For example, in a citation network, node labels correspond to the field of research of the corresponding paper or author.
- Consider the example network shown in Fig. 2.3.

  – It has 11 nodes out of which 8 of them are already labeled.

- There are 3 nodes that are colored green, three others are colored yellow and two nodes are colored blue. Nodes with the same color are from the same class.
- There are 3 nodes that are colored grey and they are unlabeled.

- In real-world data also, some of the nodes/edges/graphs are unlabelled. Hence, the task of classification aims to classify these unlabeled entities into their respective classes by making use of the existing structure and information that encompasses the network.
- If the learning process is (semi) supervised, which uses labels during training, then learned embeddings are more discriminative as they are trained in an end-to-end fashion with the end task being the classification task.
- However, in unsupervised learning, first, the embeddings are learned, and then popular classifiers, e.g. logistic regression is used with the learned embeddings forming the input features.
- For example, in Fig. 2.3, all grey colored nodes are unlabeled, and the classification process will classify these nodes into appropriate classes. We expect the grey colored node in the neighborhood of the three green colored nodes to be classified so that it is also colored green.
- Evaluation metrics for analyzing the classification performance typically are *classification accuracy* and *F1 score*.

2.6.2 Clustering

Many nodes share the same property in a network, and together they form a soft cluster or community. For instance, in a social network, users with interest in the same domain will form a community/cluster; in a citation network, papers having similar research domains form a single group. The process of finding such soft clusters is called *community detection* in the context of network analysis.

- This task aims to partition nodes or (sub)graphs into various groups in a way that similarity between samples within a cluster is maximum, whereas samples from different clusters are dissimilar.
- The partition is typically soft; it can be hard if the application demands.
- This task is very important in the biological field (proteins-proteins network, drug-disease network) to find related and interacting drugs or associated diseases.
- In the example network shown in Fig. 2.3, there are three clusters based on connectivity among the nodes; connectivity provides the similarity here. These are shown using dotted circles.
- Similarly, in a set of graphs datasets, many graphs share the same property(ies) and belong to the same cluster. For example, graphs that have a similar clique structure can be grouped.
- Any generic clustering algorithm such as K-means, K-means++, or *LDA* can be used after learning node or (sub)graph embeddings to obtain communities.

- It is quite similar to the community detection task, and if the learned representations are discriminative, they should be able to detect appropriate community structure exhibited by the graph.
- The embedding's performance on the clustering task is measured by purity, ARI, and normalized mutual index metrics by leveraging ground truth clusters based on the node labels.

2.6.3 Link Prediction (LP)

It aims to predict new connections that are likely to occur in the future based on the existing relations in the graph and the network properties.

- This will further throw light on how the network evolves.
- Link prediction can also infer the missing connections (edges) in the network.
- Major applications of link prediction are in recommender systems, social networks, biological networks. In social networks, one can use LP to predict friendship among users of the network and recommend the friendship connections. Biological networks (drug-disease networks) take advantage of this task to expand the biological dataset and infer new diseases or treatments via predicting interactions between entities.
- For example, all the dotted lines between a pair of nodes in Fig. 2.3 are the predicted links.
- For evaluation, some links (connections) from the given graph are deleted followed by learning node embeddings using the new graph. These embeddings are then used to predict these deleted edges.
- The proficiency of embeddings on how well they can support the network evolution depends on the quality measured by the metrics AUC and precision.

2.6.4 Visualization

This task helps in data mining and analyzing real-world high dimensional data visually by projecting it into two or three dimensions.

- Once low-dimensional embeddings are trained, which encode the network structure, any of the available visualization tools can be used.
- For instance, t-SNE takes as input the learned encodings and projects them into 2D or 3D space. Such an easy to visualize plot will help in inferring clusters or communities.
- Principal Components can also be used for mapping the embeddings into a lower dimension space for easy visualization.
- These visualizations provide insights into the quality of the embeddings based on whether the nodes from the same class (or belonging to the same community) are

close to each other. Also, nodes with different class labels need to be far away in the projected space.

2.6.5 Network Reconstruction

- It aims at reconstructing the actual network (graph) using similarity between each pair of node embeddings.
- The similarity score between embedding vectors determines the similarity between two nodes and is used to infer the edge between them.
- If the learned embeddings are discriminative enough, then the underlying similarity function should detect the original links present in the network, which determines the quality of these learned representations.
- In the process, it will be interesting to look at existing edges that need to be deleted based on the low similarity between the embeddings of the end vertices.

2.7 Embeddings Based on Matrix Factorization

We have seen that a popular way of representing a network graph $G = (V, E)$ is to use its adjacency matrix A.

- The $N \times N$ matrix A captures the structure/connectivity information. Each row of the matrix is a vector of size N based on the presence/absence of links of a node to all the nodes in the network. In a practical setting N could be very large.
- In addition to the connectivity, if each node has some content/attribute information, then a matrix X of size $N \times D$ is used. Here each node is viewed as a vector of size D, where D is the size of the vocabulary behind the content, which could also be large.
- So, we need to represent both the structure and content data in a low-dimensional space to facilitate efficient and accurate machine learning on the network data.
- *Matrix factorization* is one of the well-known techniques to reduce the dimensionality. This may be viewed as follows:
  - We can factorize the adjacency matrix A as $A = BC$ where A is a $N \times N$ adjacency matrix, B is a $N \times K$ matrix, and C is a $K \times N$ matrix.
  - Similarly the matrix X can be factorized as $X = GH$ where X is of size $N \times D$, G is a matrix of size $N \times K$, and H is a matrix of size $K \times D$.
  - Typically the value of K is much smaller than both N and D to facilitate dimensionality reduction.
  - If the structure and content in the network are in perfect agreement with each other, then their low-dimensional representations will also be agreeing with each other. This could be to the extent that the two matrices B and U are equal which can simplify the factorizations. In practice, B and U could be different because

of noise in the correspondence. In such a case one can aim to minimize some difference between the two matrices.

There are several well-known techniques for factorization of matrices. These are considered in the next few subsections.

2.7.1 Singular Value Decomposition (SVD)

Singular value decomposition (SVD) is the most popular matrix factorization technique. Either a square matrix ($N \times N$ matrix) like A or a rectangular matrix ($N \times D$ matrix) like X can be factorized using SVD.

- In general any real matrix S of size $p \times q$ can be decomposed into

$$S_{p \times q} = U_{p \times K} \Sigma_{K \times K} V^T_{K \times q}, \ where$$

  – V^T is the transpose of the matrix $V_{q \times K}$.
  – The non-zero eigenvalues of SS^T and $S^T S$ are the same; it is possible that some of these eigenvalues are repeated. If $p > q$ then SS^T will have at least $(p - q)$ 0 (zero) eigenvalues. Similarly, if $q > p$ then $S^T S$ will have at least $(q - p)$ 0 (zero) eigenvalues. The common eigenvalues are non-negative.
  – Σ is a diagonal matrix of size $K \times K$. Its diagonal entries are the positive square roots of the K largest eigenvalues of either SS^T or $S^T S$, where $K < min(p, q)$.
  – These diagonal entries of Σ are called the *singular values* of S. They are non-negative real numbers, if S is real, and are typically arranged in non-increasing order.
  – Columns of U are the K eigenvectors of SS^T and columns of V are the K eigenvectors of $S^T S$. These K vectors are the orthonormal eigenvectors associated with the top K common eigenvalues.

- So, SVD can be used for matrix factorization. Further, it can be used to obtain the *principal components* (PCs) of the data present in the form of the row vectors of either A or X matrices associated with the graph representing any network.
- If the rows of the matrix S are normalized so that their mean is zero, then the eigenvectors of $S^T S$ (columns of V) are the principal components.
- If the matrix to be factorized is symmetric, then we can decompose it using orthogonal matrices. For example, the adjacency matrix A will be a symmetric matrix if the graph is undirected. In such a case we have

$$A_{N \times N} = P_{N \times K} D_{K \times K} P^T_{K \times N},$$

where P is an orthogonal matrix, that is $P^T = P^{-1}$.

2.7.2 Matrix Factorization Based Clustering

It is possible that some of the entries in U or V or both can be negative. A consequence of this is that even the PCs can have negative entries.

- We can illustrate this using a simple example. Consider the dataset shown in Table 2.6.

  - There are four objects represented as two-dimensional patterns in the table; each is described by the Volume of the object and its Price both in respective units. For example, the first object has 1 unit of volume and 8 units of price.
  - The sample mean of the 4 points is (2,5). So, the zero mean normalized data is given by the matrix

  $$Z = \begin{bmatrix} -1 & 3 \\ 1 & -3 \\ -1 & 3 \\ 1 & -3 \end{bmatrix}$$

  - The matrix $Z^T Z$ is given by

  $$Z^T Z = \begin{bmatrix} 4 & -12 \\ -12 & 36 \end{bmatrix}$$

  - The eigenvalues of the matrix $Z^T Z$ are 40 and 0 and its orthonormal eigenvectors are $\begin{pmatrix} \frac{1}{\sqrt{10}} \\ \frac{-3}{\sqrt{10}} \end{pmatrix}$ and $\begin{pmatrix} \frac{3}{\sqrt{10}} \\ \frac{1}{\sqrt{10}} \end{pmatrix}$.
  - These two vectors are the two PCs in that order and they are orthogonal to each other and there is a negative entry in the first PC.
  - This example clearly illustrates that PCs can have negative entries.

- However, there could be applications where we require only non-negative entries in the factor matrices. Clustering or community detection is one such example. We can explain using the example data in Table 2.6.

  - Note that in the two-dimensional space characterized by Volume and Price, pattern 1 and pattern 3 are identical. Similarly, pattern 2 and pattern 4 are also identical.
  - So, if we want to assign these 4 patterns into two clusters, then pattern 1 and pattern 3 are in one cluster and the remaining two patterns are in the other cluster.
  - Such a cluster structure may be realized using the matrix factorization, $X_1 = G_1 H_1$, that is exemplified by

  $$\begin{bmatrix} 1 & 8 \\ 3 & 2 \\ 1 & 8 \\ 3 & 2 \end{bmatrix} = \begin{bmatrix} 1 & 0 \\ 0 & 1 \\ 1 & 0 \\ 0 & 1 \end{bmatrix} \begin{bmatrix} 1 & 8 \\ 3 & 2 \end{bmatrix}$$

- Observe that G_1 matrix (the 4×2 matrix in the RHS) is the *cluster assignment matrix*. The two columns of G_1 correspond to the two clusters. Pattern 1 and pattern 3 are assigned to cluster 1 and the corresponding entries in column 1 are 1. Similarly, pattern 2 and pattern 4 are assigned to cluster 2 and the second column in matrix G_1 indicates this assignment.
- It is important to note that the entries in G_1 and H_1 are nonnegative. This property is essential here.
- Note further that the matrix H_1 is the *cluster description matrix*. The two rows of H_1 describe the two clusters. Here, each cluster is described by its centroid.
- In this simple example, both the patterns in each cluster are identical. In general, a centroid-based clustering algorithm like the K-means clustering algorithm describes each cluster by its centroid. In such a case, the ith row of matrix H_1 will be the centroid of the ith cluster, for $i = 1, \ldots, K$.
- For example, if pattern 3 and pattern 4 in Table 2.6 are changed to $(1,6)$ and $(5,2)$ respectively, then the factorization, for $X_2 \approx G_2 H_2$ is given by

$$\begin{bmatrix} 1 & 8 \\ 3 & 2 \\ 1 & 6 \\ 5 & 2 \end{bmatrix} \approx \begin{bmatrix} 1 & 0 \\ 0 & 1 \\ 1 & 0 \\ 0 & 1 \end{bmatrix} \begin{bmatrix} 1 & 7 \\ 4 & 2 \end{bmatrix}$$

where X_2 is the modified version of X_1 and G_2 and H_2 are respective cluster assignment and cluster description matrices.
- Note that pattern 1 and pattern 3 are assigned to cluster 1 and the remaining two patterns are assigned to cluster 2. In this case, $G_2 = G_1$ and H_2 has the cluster centroids $(1,7)$ and $(4,2)$ as its rows.
- Observe further that the factorization of X_1 is exact whereas the factorization of X_2 is approximate.
- In general, any hard clustering output can be abstracted by such a non-negative matrix factorization where $X \approx GH$ where each row of G will have one entry 1 and the remaining $K - 1$ entries 0.
- If the entry G_{ij} in the ith row and the jth column of G is 1, then it indicates that the ith pattern, that is the ith row of X is assigned to cluster j.
- All other entries in the ith row of G will be 0; that is $G_{ik} = 0, \ \forall k \neq j$. This is because in hard clustering a pattern is assigned to one and only one cluster.
- Further, $H(i)$, the ith row of H represents the ith cluster. It could be the centroid in the case of K-means algorithm; but in general it could any vector that is a suitable *representative of the ith cluster*.

Every *hard clustering* output can be viewed as leading to such a matrix factorization. For example, the output of spectral clustering also can be represented in terms of matrix factorization.

Table 2.6 Example data matrix

Pattern	Volume	Price
1	1	8
2	3	2
3	1	8
4	3	2

2.7.3 Soft Clustering as Matrix Factorization

There are several applications where a natural requirement is to assign a pattern to more than one cluster. For example, a document may share more than one topic/cluster; for example, it may be dealing with both *sports* and *politics*. In such a case, we require the clustering to be soft.

- *Topic models* like the latent Dirichlet allocation (LDA) are probabilistic and they assign a document to more than one topic/cluster.
- In such a case, in the approximation of X as GH, the ith row of G has one or more non-zero entries. The entries in any row of G add upto 1. G_{ij} could be viewed as the probability that the ith document belongs to the jth topic/cluster.
- The ith row of H is the probabilistic description of the ith topic/cluster. In fact a topic is an assignment of a probability value to each term in the vocabulary present in the collection of documents.
- It is not just the LDA alone. Every soft clustering output could be represented using an appropriate factorization of matrices.
- For example, probabilistic latent semantic indexing ($PLSI$) is one such example. The output of $PLSI$ can be abstracted as

$$S_{p \times q} = U_{p \times K} Z_{K \times K} V_{K \times q}$$

- Here the $p \times q$ matrix S represents p documents using q vocabulary terms, where the input is a collection of p documents and the vocabulary size is q.
- The matrix U is of size $p \times K$. It describes p documents using K topics/soft clusters.
- The matrix V of size $K \times q$ is a description of the K topics using the q vocabulary terms. The ith row of V describes the ith topic for $i = 1, \ldots, K$ and each row is a q-dimensional vector with one or more probability entries.
- The $K \times K$ matrix Z is a diagonal matrix that describes the strength of each of the topics. The diagonal entry in the ith row and the ith column of Z, that is Z_{ii} indicates the importance of topic i in the collection.

There could be applications where the factorization could be deterministic rather than probabilistic.

2.7.4 Non-Negative Matrix Factorization (NMF)

One of the well-known matrix factorization approaches that is deterministic is the NMF. Here, a matrix X is approximated using two factors G and H.

- It is viewed as minimizing the squared euclidean norm between X and GH. So, It is based on
$$||X - GH||^2.$$

- This norm is the sum of element-wise deviations or the squared deviations between X_{ij} and $(GH)_{ij}$ are added over all the elements, that is for $i = 1, \ldots, N$ and $j = 1, \ldots D$.
- We find G and H so that $||X - GH||^2$ is minimized under the constraints that $G_{ij} \geq 0$ and $H_{ij} \geq 0$ for all i and j.
- If X and G are known then finding H is a well-behaved convex optimization problem. Similarly, if X and H are given, then finding G is also a convex optimization problem.
- So, the problem of finding G and H given X is solved by using an *alternating minimization* process. Using X and some initial G, H is computed. Using X and the obtained H, G is updated. This process goes on till some termination condition is satisfied.
- The solution obtained by this optimization is locally optimal. Note that several of the factorization schemes can give us only a local optimum. These include the K-means clustering, LDA, $PLSI$, and NMF.
- On the other hand SVD and the orthogonal decomposition are deterministic and give us the exact factorization.
- The motivation behind NMF was that the factors G and H provide information about the presence/absence of parts of objects in the data.
- Consider the example matrix factorization given by

$$\begin{bmatrix} 1 & 1 & 0 & 0 & 1 & 1 \\ 1 & 1 & 0 & 0 & 1 & 1 \\ 0 & 0 & 1 & 1 & 0 & 0 \\ 0 & 0 & 1 & 1 & 0 & 0 \end{bmatrix} = \begin{bmatrix} 1 & 0 \\ 1 & 0 \\ 0 & 1 \\ 0 & 1 \end{bmatrix} \begin{bmatrix} 1 & 1 & 0 & 0 & 1 & 1 \\ 0 & 0 & 1 & 1 & 0 & 0 \end{bmatrix}$$

- This illustrates NMF. Note that the 4×6 matrix in the LHS is a rank 2 matrix. It clearly shows two linearly independent row vectors, (110011) and (001100) and the two linearly independent column vectors $\begin{pmatrix} 1 \\ 1 \\ 0 \\ 0 \end{pmatrix}$ and $\begin{pmatrix} 0 \\ 0 \\ 1 \\ 1 \end{pmatrix}$.
- The factorization aptly captures this by exploiting the fact that the rank of the matrix is 2 ($K = 2$).
- In general, the value of K is upper bounded by the minimum of N and D where X is of size $N \times D$.

- Further note that the independent columns form the G matrix and the independent rows form the H matrix. These basis vectors may be viewed as parts of the objects or patterns present in X.
- Further, note that there are two hard clusters and the assignment of the first two vectors to the first cluster and the remaining two rows to the second cluster are depicted by the 4×2 matrix in the example.
- Similarly, the two rows of the 2×6 matrix, in the RHS of the example, capture the description of the two clusters.
- A fundamental observation from linear algebra is that *row rank = column rank = rank* of any matrix.
- This implies that clustering and dimensionality reduction have an excellent correspondence as depicted in the example factorization.

Even though matrix factorization is very popular in several machine learning applications, its main drawback is that it is computationally not very attractive to deal with large-scale network datasets. An alternative is based on *random walks*. We consider in the next section, *word2vec* that offered the basic platform for several state-of-the-art network embedding schemes that use random walks.

2.8 Word2Vec

It is the most influential tool in terms of its impact on network embedding schemes. It deals with representing each word as a vector. Neural network models require real-valued vectors as input. So, a string or a word needs to be converted into a vector of numbers to be processed by a neural network model to carry out various tasks. Word2vec is one of the popular techniques, and it is described below:

- Word2Vec employs a neural network model with a single hidden layer. It generates real-valued vector representations, called neural word embeddings, for all the words in the vocabulary.
- These word embeddings provide a means to calculate the similarity between words, sentences, and consequently, the documents.
- First step is to create a training corpus as follows:

  – All unique words will make up the vocabulary (W), and each unique word will be given an index between 1 and |W|, where |W| is the number of unique words.
  – Word2vec uses words and their contexts, based on the principle that words with similar context are similar.

    · *Context* of a word is a set of all those words which occur within a window of fixed size in a sentence with the focused word in the middle.
    · Formally, the context of a word with window size fixed to ws includes ws words before and ws words after the focused word, totaling into $2ws$ words.

· For example, let $w_1, w_2, ..., w_{i-2}, w_{i-1}, w_i, w_{i+1}, w_{i+2}, ...w_{n-1}, w_{n-2}$ be a sentence, then the context of word w_i for the window of size 2 is $\{w_{i-2}, w_{i-1}, w_{i+1}, w_{i+2}\}$.

· The training corpus contains all (word, context) pairs for all the words in the vocabulary, such as for the word w_i: $(w_i, w_{i-2}), (w_i, w_{i-1}), (w_i, w_{i+1})$, (w_i, w_{i+2}) word-context pairs are in the training set.

- Given the corpus, the word embeddings can be learnt using two different schemes.

   1. *CBOW* (continuous bag of words) takes the context words as input and predicts the probability of the target word corresponding to the input.
   2. *Skip-Gram* works by taking the word as input and predicting the words in its context. So, from each pair in the training corpus, it takes the word as input and predicts the probability distributions corresponding to the context words.

- The learned encodings have some implicit dependencies, unlike one-hot vectors. One hot vector is a binary vector of size $|W|$, which will have 0s in all positions but 1 in one location that indexes the word being encoded.

- In the following section, we will explain the skip-gram model in more detail because skip-gram gives more precise results than CBOW even while using smaller data sets.

2.8.1   Skip-Gram Model

Figure 2.4 shows the architecture of the skip-gram approach. Some important points to observe from the figure and the working of the skip-gram model are summarized below.

- U_1 is a weight matrix of dimension $|W| \times d$ (where d is the dimension of the word representations) between input and the hidden layer. U_2 is a weight matrix of dimension $d \times |W|$ between hidden and the output layers.

- Multiplication of the input vector with U_1 gives a row of U_1 corresponding to the input. This is the vector of the hidden layer and has a dimension of $1 \times d$. The output vector of dimension $1 \times |W|$ is computed by multiplying this hidden layer vector with U_2. According to the window size (i.e., the number of the context words), this output vector is repeated.

- In the figure, it takes one-hot vector of the word as input and generates two vectors (VECTOR1 and VECTOR2) of probabilities. The ith index of the vector denotes the probability of the ith vocabulary word being the input's context word.

In general, a skip-gram model generates $2.ws$ output vectors, one for each context word, where ws is the window size. After training, the matrix U_1 is the required word representation i.e., each row of the first layer weight matrix is interpreted as the word embedding of the corresponding word in the vocabulary. Thus the goal of word2vec is just to train this weight matrix of the hidden layer and each word will be represented

Fig. 2.4 Word2vec: a
skip-gram model

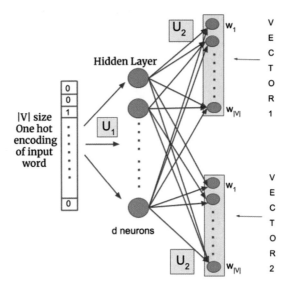

by a vector of size $1 \times d$, where d is the dimension and the number of nodes in
the hidden layer. The main objective of word2vec is to output similar probability
distributions for the words with similar contexts, and consequently, those words
should have similar embeddings too. Precisely, the model outputs the probability of
being the context word of the input for all the words.

   Moreover, when the model is trained on the entire training set, all the weights
of the model are modified for each pair in the corpus. So, on a large corpus, the
computation becomes challenging to track. Therefore, *negative sampling* is used to
train the word2vec model.

- Instead of updating all the weights of the model for each training pair, it will update
  the parameters corresponding to the positive word along with K negative words.
- Formally saying, for each positive (word, context) pair, it will sample k negative
  (word, context) pairs and not the complete training set.
- These pairs are called negative because the pair's context word is randomly chosen,
  not from nearby words.
- Loss is propagated for these selected K entries. The corresponding weights are
  updated along with the weights of the positive word by maximizing the log-
  likelihood corresponding to the positive sample and minimizing the log-likelihood
  of the selected negative pairs.

2.9   Learning Network Embeddings

Any network embedding method generates vectors of low dimension representing
an entity of the network such as a node, an edge, or the entire (sub)graph. These are
called node embedding, edge embedding, or whole (sub)graph embedding, respectively. Different types of embeddings have diverse applications such as whole graph
embedding can facilitate the grouping of multiple graphs together. In contrast, node
embeddings are used for node classification, node clustering, and edge prediction
tasks.

- Node and Edge Embeddings:

  1. The aim of node embedding is to represent each node of the network as a low-
     dimensional vector.

     – They preserve relations between the nodes of the network in the form of
       geometric relations between node embeddings.
     – Each node representation learning approach aims at preserving different
       properties of the network.
     – So, existing approaches differ in how they encode the similarity between
       nodes and what node similarity the approach accounts for.
     – For example, some approaches preserve the macroscopic structure that cap-
       tures scale-free properties. Some maintain first or second-order proximities
       between nodes, and other methods embed nodes based on their roles in the
       network.
     – These embeddings are then exploited by many machine learning downstream
       tasks such as classification and clustering of nodes.
     – Chapter 4 deals with the topic of node embeddings.

  2. Edge embedding aims to encode edges of the network in a vector space.

     – Edges encode node pair relations (pairwise similarity between nodes).
     – Major challenges involve dealing with asymmetry, calculating edge level
       similarities to encode edge semantics, and facilitating various edge-based
       tasks such as link prediction.

- Graph Embedding: It aims at representing a set of nodes or an entire graph in a
  low-dimensional vector space.

  – Embeddings are such that the properties of the entire graph are captured using
    similarity between graphs, i.e., the algorithm generates a single vector for the
    whole (sub)graph.
  – An important requirement is to keep similar graphs close in the embedding
    space.
  – These embeddings have many critical applications, from predicting the class
    label of an entire graph (graph classification) to clustering graphs.
  – For instance, finding anti-cancer activity, finding molecule toxicity level, and
    many more can be tackled by embedding the entire graph.

- Some algorithms first generate embeddings for nodes using node representation learning algorithms and then use any aggregation operator (mean, average, maximum) on those embeddings to output a single vector.
- A myriad of graph pooling operations is recently developed, which output the graph's coarsened versions and finally represents the entire graph with a single vector.
- A detailed discussion of these approaches is provided in Chap. 6.

Also, we can group the embedding learning algorithms into Supervised and Unsupervised, as explained below:

- Supervised Learning:

  - Supervised or semi-supervised learning of embeddings depends primarily on label information to learn the model's parameters as the loss is controlled by actual and predicted labels.
  - During the training stage, class label information and information in matrices A (structure) and X (attribute) are exploited. Therefore more discriminative representation learning takes place.
  - Though a small proportion of nodes are labeled, labels proffer information about the categorization of network entities.
  - The learning in this setting depends on the downstream tasks such as classification or link prediction; therefore, these embeddings may be limited to be used for a particular task only.
  - Further, collecting labels is very expensive, which is another disadvantage of these approaches.

- Unsupervised Learning:

  - These approaches can work without the label information while learning representations.
  - So, these methods can be used to learn embeddings even when class label information is not available.
  - Unsupervised learning algorithms use structural information and sometimes attribute information (if available) during the training stage.
  - Since unsupervised algorithms do not exploit any downstream task knowledge for embedding learning; therefore, the embeddings learned are generic and task agnostic, unlike embeddings learned by supervised approaches.
  - An advantage of these embeddings is that they can be used across all downstream tasks for further analysis, such as classification and clustering.

2.10 Summary

- We provided a brief introduction to network representation.
- Graphs are popularly used in representing networks.
- Adjacency matrix associated with the network graph captures the structural properties of the network.
- Content matrix captures the attribute information or content associated with the nodes in the network.
- matrix factorization plays an important role in arriving at a low-dimensional representation of the network entities and even the entire network.
- The low-dimensional embeddings obtained are useful in dealing with several ML tasks including classification, community detection, link prediction, visualization and network construction.
- Word2vec is the most influential tool in network embedding. Random walk over nodes in the graph can be viewed as some kind of sentences.
- High degree nodes will appear in more such walks. The sequences of vertices visited can be viewed like sentences where each node in the walk can be viewed like a word.
- So word2vec can be used in the network/graph applications to embed nodes, where a random walk over vertices is like a sentence and a vertex is like a word in the sentence.
- These ML tasks are instrumental in several important application domains. Some examples are:

  - *E-commerce*: Recommending one or more products to customers
  - *Education*: Finding concepts that are not well understood by a student, interrelationships between different concepts/topics
  - *Health*: Identifying possible drug interactions
  - *Transportation*: Suggesting optimal routes
  - *Agriculture*: Estimating crop yields
  - *Security*: Locating terrorist groups

Bibliography

1. Hamilton WL, Ying R, Leskovec J (2017) Representation learning on graphs: methods and applications. arXiv preprint arXiv:1709.05584
2. Goyal P, Ferrara E (2018) Graph embedding techniques, applications, and performance: a survey. Knowl Based Syst 151:78–94
3. Cui P, Wang X, Pei J, Zhu W (2018) A survey on network embedding. IEEE Trans Knowl Data Eng 31(5):833–852
4. Zhang D, Yin J, Zhu X, Zhang C (2018) Network representation learning: a survey. IEEE Trans Big Data

5. Debnath AK, Lopez de Compadre RL, Debnath G, Shusterman AJ, Hansch C (1991) Structure-activity relationship of mutagenic aromatic and heteroaromatic nitro compounds. Correlation with molecular orbital energies and hydrophobicity. J Med Chem 34(2):786–797. http://graphkernels.cs.tu-dortmund.de

6. McCallum AK, Nigam K, Rennie J, Seymore K (2000) Automating the construction of internet portals with machine learning. Inf Retr 3(2):127–163

7. Dobson PD, Doig AJ (2003) Distinguishing enzyme structures from non-enzymes without alignments. J Mol Biol 330(4):771–783. http://graphkernels.cs.tu-dortmund.de

8. Schomburg I, Chang A, Ebeling C, Gremse M, Heldt C, Huhn G, Schomburg D (2004) BRENDA, the enzyme database: updates and major new developments. Nucleic Acids Res 32(suppl_1):D431–D433. http://graphkernels.cs.tu-dortmund.de

9. Breitkreutz BJ, Stark C, Reguly T, Boucher L, Breitkreutz A, Livstone M, Oughtred R, Lackner DH, Bähler J, Wood V, Dolinski K (2007) The BioGRID interaction database: 2008 update. Nucleic Acids Res 36(suppl_1):D637–D640

10. Leskovec J, Kleinberg J, Faloutsos C (2007) Graph evolution: densification and shrinking diameters. ACM Trans Knowl Disc Data (TKDD) 1(1):2-es

11. Tang J, Zhang J, Yao L, Li J, Zhang L, Su Z (2008) August. Arnetminer: extraction and mining of academic social networks. In: Proceedings of the 14th ACM SIGKDD international conference on knowledge discovery and data mining, pp 990–998

12. Tang L, Liu H (2009) Relational learning via latent social dimensions. In: Proceedings of the 15th ACM SIGKDD international conference on Knowledge discovery and data mining, pp 817–826

13. Tang L, Liu H (2009) Scalable learning of collective behavior based on sparse social dimensions. In: Proceedings of the 18th ACM conference on information and knowledge management, pp 1107–1116

14. Kersting K, Kriege NM, Morris C, Mutzel P, Neumann M (2016) Benchmark data sets for graph kernels. http://graphkernels.cs.tu-dortmund.de

15. Lee DD, Seung HS (2001) Algorithms for non-negative matrix factorization. In: Advances in neural information processing systems, pp 556–562

16. Koren Y, Bell R, Volinsky C (2009) Matrix factorization techniques for recommender systems. Computer 42(8):30–37

17. McCormick C (2016) Word2Vec tutorial - the skip-gram model. http://mccormickml.com/2016/04/19/word2vec-tutorial-the-skip-gram-model/

18. McCormick C (2017) Word2Vec tutorial part 2 - negative sampling. http://mccormickml.com/2017/01/11/word2vec-tutorial-part-2-negative-sampling/

19. NSS (2017) An intuitive understanding of word embeddings: from count vectors to Word2Vec. https://www.analyticsvidhya.com/blog/2017/06/word-embeddings-count-word2veec/

Chapter 3
Deep Learning

3.1 Introduction

Representation is the most fundamental issue in network analysis. More generically it affects the performance of any machine learning system. For example *weight of objects* alone is adequate to classify objects into *lighter* and *heavier* classes. Similarly, *height* is adequate to discriminate objects into *tall* and *short* classes. Such choices are simple in real life and are often based on commonsense.

However, in most of the practical applications, it is not possible to come out with a representation of the data so easily. However, a good representation is essential for successful machine learning. This may be attributed to the raise in the usage of *deep learning* systems. A routine way of appreciating deep learning is that the underlying learning system is realized using a cascade of systems that successively process data and pass on the information to subsequent levels; the size of the cascade is an indication of the *depth* of the learning system.

A hallmark of a deep learning systems is:

- **Representation Learning:**
  Can the system learn the representation automatically from the given data?
  In order to answer this question, we need to pose additional questions like:

  1. What is the size of the data required to learn the *right representation* automatically?
  2. What is the type of data that can be processed?
  3. Is it required to scale/normalize the data?
  4. Will the performance be affected by the order in which the data is processed?
  5. Is the model learnt for one application generic enough to be used in other applications?

Even though it is possible for a variety of realizations to answer one or more of these questions convincingly, it is the *artificial neural network* based systems that are

M. Aggarwal and M. N. Murty, *Machine Learning in Social Networks*,
SpringerBriefs in Computational Intelligence,
https://doi.org/10.1007/978-981-33-4022-0_3

shown to answer most of these questions. So, they are the most popular and perhaps only systems available for deep learning currently. So, it is convenient to view deep learning and deep neural networks as synonymous; we take this stand in rest of the chapter.

3.2 Neural Networks

Artificial neural networks ($ANNs$) are used by default for deep learning. Before we go into an exposition of deep neural networks, we will examine the basic building blocks that are essential in understanding the functionality of deep neural networks in this section. Historically, there were several developments in the early days but one of the simplest and important milestones was perceptron. We examine it and then consider more deeper architectures.

3.2.1 Perceptron

The working of Perceptron may be explained using Fig. 3.1

- Let X be an l-dimensional vector corresponding to a train or a test pattern. Such a pattern is given by $X = \begin{pmatrix} x_1 \\ x_2 \\ \vdots \\ x_l \end{pmatrix}$.

- Let $\phi_i(X)$, $i = 1, \ldots, d$ be features extracted from X. So, $\phi_i s$ are mappings from $\Re^l$ to $\Re$. For example, if X is a 2-dimensional vector given by $X = \begin{pmatrix} x_1 \\ x_2 \end{pmatrix}$, then $\phi_1(X) = x_1$, $\phi_2(X) = x_2$, and $\phi_3(X) = x_1 x_2$ is a possible set of 3 features. Here, $l = 2$ and $d = 3$.

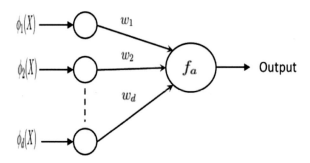

Fig. 3.1 Perceptron in the feature space

- The *weights* w_1, w_2, ..., w_d indicate the importance of $\phi_1(X)$, $\phi_2(X)$, ..., $\phi_d(X)$ respectively. We call the perceptron using such a generic representation as a *perceptron in the feature space*
- The unit indicated by f_a is called the *activation unit* where f_a is the *activation function*. The sum of weighted features given by $\sum_{i=1}^{d} {}_i\phi_i(X)$ is the input to f_a; f_a is a function from $\Re$ to $\Re$. A simple example of f_a is

$$f_a(\alpha) = \begin{cases} 1 & if \quad \alpha > 0 \\ 0 & otherwise \end{cases} \tag{3.1}$$

Such an activation function f_a is called a *Linear Threshold Function*.
- Note that $Output = f_a(\sum_{i=1}^{d} w_i\phi_i(X))$ which in general is a *nonlinear function* of the weighted sum $\sum_{i=1}^{d} w_i\phi_i(X))$.
- Let us consider a simple example to illustrate its working:
  Example: Consider the following five 2-dimensional patterns.

Negative Class: $\begin{pmatrix} 1 \\ 2 \end{pmatrix}$, $\begin{pmatrix} 2 \\ 3 \end{pmatrix}$

Positive Class: $\begin{pmatrix} 3 \\ 1 \end{pmatrix}$, $\begin{pmatrix} 4 \\ 1 \end{pmatrix}$, $\begin{pmatrix} 4 \\ 2 \end{pmatrix}$

– Note that we have $l = 2$ in this example. Further let us assume that $\phi_1(X) = x_1$ and $\phi_2(X) = x_2$. So, in this case $l = d = 2$ and the input features are the features used.
– Using some algorithm suppose we *learn the weights* to be $w_1 = 1$ and $w_2 = -1$. So the weighted sum is given by $\sum_{i=1}^{2} w_i\phi_i(X) = x_1 - x_2$.
– If we use the linear threshold function f_a on the weighted sum, we make the following decision:
  for a pattern $X = \begin{pmatrix} x_1 \\ x_2 \end{pmatrix}$, if $x_1 - x_2 > 0$, then classify X as a positive class pattern and if $x_1 - x_2 < 0$ then assign X to the negative class.
– Note that for pattern $\begin{pmatrix} 1 \\ 2 \end{pmatrix}$, $x_1 - x_2 = -1 < 0$; So, it is classified as a member of the negative class. Similarly, for $\begin{pmatrix} 3 \\ 1 \end{pmatrix}$, $x_1 - x_2 = 2 > 0$; So, it is assigned to the positive class. Further, $x_1 - x_2 = 0$ characterizes the boundary to decide between the two classes.
– We depict the example using Fig. 3.2. In the figure, the two patterns of the negative class and the three patterns of the positive class are shown. Further, the dotted line $x_1 = x_2$ or equivalently $x_1 - x_2 = 0$ is the *decision boundary* between the two classes characterized by $w_1 = 1$ *and* $w_2 = -1$.
– There is another line, a broken line, which is parallel to the earlier line and is described by $x_1 = x_2 + 1$. Note that even this line also is a decision boundary between the two classes. It is possible to see that if there exists one decision boundary, there can be infinite decision boundaries between the two classes.

Fig. 3.2 Decision boundaries

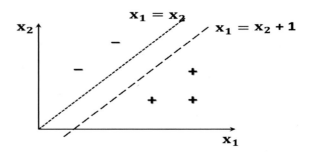

- The decision boundary $x_1 = x_2 + 1$ or equivalently $x_1 - x_2 - 1 = 0$ may be explained by a different choice of $\phi_i s$ and $w_i s$. If we define $\phi_0(X) = 1$ and $w_0 = -1$ and retain the earlier predicates and weights, then the decision boundary is described by $\sum_{i=0}^{2} w_i \phi_i(X) = -1 + x_1 - x_2 = 0$. This is a more generic form that relaxes the constraint that the decision boundary goes through the origin.

- In the two-dimensional example we have considered the decision boundary to be a line and its generic form is $\sum_{i=0}^{2} w_i \phi_i = 0$. These ideas can be extended to deal with binary classification (two-class) problems in any l dimensions by using $\sum_{i=0}^{d} w_i \phi_i$ as the weighted sum or input to the activation function f_a. In such a case, the decision boundary is a hyperplane.

- A popular choice for the features is $\phi_0(X) = 1$ and $\phi_i(X) = x_i$ for $i = 1, 2, \ldots, l$. The advantage of this representation is that it requires $d + 1$ features (including ϕ_0) where $d = l$. So, the complexity is linear in the input dimension l. Let us call a perceptron using such a representation as a *perceptron in the input space* However, the resulting decision boundary cannot deal with two classes of patterns that are not linearly separable.

- An important point is that by considering a larger value of d, it is possible to deal with nonlinear classification problems. Specifically, when the patterns are l-bit binary strings, it is possible to represent any boolean function on l bits using all possible subsets (minterms of different sizes) as features.

- For example, the function $odd - parity(x_1, x_2, x_3)$ returns a 1 if $x_1 + x_2 + x_3$ is odd and returns a value 0 otherwise. It is not linear in terms of the four features $\phi_0(X) = 1$ and $\phi_i(X) = x_i$ for $i = 1, 2, 3$ where X is a 3-bit binary pattern. However by using additional features, or minterms, it is possible to represent the odd-parity function using the form $x_1 + x_2 + x_3 - 2(x_1x_2 + x_1x_3 + x_2x_3) + 4x_1x_2x_3$.. Note that such a representation involves features that are nonlinear in $x_1, x_2,$ *and* x_3 like $x_1x_2, x_1x_3, x_2x_3,$ and $x_1x_2x_3$.

- Any vector can be represented as a binary string of some length l on a boolean computer. So, any classification problem based on training patterns can be dealt with a *perceptron in the feature space* that employs 2^l features. However, in most of the practical problems the value of l, the number of bits could be very large and training a perceptron using 2^l features could be *computationally prohibitive*. That is the reason for employing a *perceptron in the input space*.

3.2.2 Characteristics of Neural Networks

Some of the important characteristics of ANNs which are related to the discussion so far are:

- They may be viewed as linear classifiers. They can handle even non-linear classification problems using an appropriate representation.
- If the classes are linearly separable in the input (l-dimensional) space, then the learning algorithms behind units like *perceptron in the input space* can find one out of the infinite possible linear decision boundaries in the $(l + 1)$-dimensional space. Popularly, perceptron in the input space is called perceptron and henceforth we too will follow this popular terminology.
- There are other linear Classifiers including the ones based on support vector machines (SVMs). An SVM constrains the search space for the decision boundary by specifying an appropriate objective function. It is possible to view an SVM also as an ANN.
- It is possible to choose an appropriate activation function to realize the associated/pre-specified nonlinearity.
- The notion of weighted sum that is used as the input of an activation function naturally imposes a constraint on the type of data that can be processed. ANNs are intrinsically capable of processing only numeric data unlike some other classifiers including the ones based on decision trees and Bayes decision theory.
- Even though some of the ANNs including SVMs normalize the data as a processing step, in theory *normalization* is not required in using ANNs. For example, if a component $\phi_i(X)$ is more important than another component $\phi_j(X)$, then the associated weight w_i can be chosen to be larger than w_j.

3.2.3 Multilayer Perceptron Networks

A perceptron cannot handle classes that are not linearly separable. Further, to get the right representation is difficult. A Multilayer perceptron (MLP) is a feedforward network that combines multiple layers, where each layer may have multiple perceptrons. A major advantage of such a network is that it has the potential to learn the required representation from the input data. As an example, consider the exclusive or (XOR) function given in Table 3.1:

Table 3.1 is the truth table of the boolean function exclusive or (XOR). The output is 1 when exactly one of the inputs is 1, but not both. The first three columns characterize the truth table. The fourth and fifth columns in the table show equivalent representations of the XOR function.

Table 3.1 Exclusive or representations

x_1	x_2	$x_1 \oplus x_2$	$x_1 + x_2 - 2x_1x_2$	$\overline{x_1}x_2 + x_1\overline{x_2}$
0	0	0	0	0
0	1	1	1	1
1	0	1	1	1
1	1	0	0	0

A perceptron cannot represent it in terms of inputs x_1 and x_2 alone as it is not a linear function in these inputs. However the fourth column suggests that if we use an additional feature x_1x_2 then it can be realized. Similarly the fifth column gives an equivalent representation using the features $\overline{x_1}x_2$ and $x_1\overline{x_2}$. These two representations can be represented using the following MLPs.

1. Representing XOR as $x_1 + x_2 - 2x_1x_2$: The corresponding MLP is shown in Fig. 3.3.

   - Note that in the figure there are three layers.
   - The *input layer* receives the inputs x_1 and x_2; note that each pattern is a two-dimensional vector here. The input layer is indicated by the presence of small circles.
   - There are two additional layers. In the output there is a perceptron whose output is the exclusive or of x_1 and x_2 shown as $x_1 \oplus x_2$. It is equivalent to $x_1 + x_2 - 2x_1x_2$.
   - There is a perceptron in the middle layer; it is also called the *hidden layer*.
   - The hidden layer perceptron outputs the AND of the inputs x_1 and x_2, that is $x_1 \wedge x_2$. It is characterized by the equivalent form $x_1 + x_2 > 1$; so it outputs 1 only when both x_1 and x_2 are 1, else a 0 (zero) exactly like an AND gate.
   - The perceptron in the output layer has 3 inputs; they are x_1, x_2 and $-2x_1x_2$. It outputs the *sum* of the three inputs giving the equivalent $x_1 + x_2 - 2x_1x_2$ of the *Exclusive OR* of x_1 and x_2 and is represented by $x_1 \oplus x_2$

2. Considering the other representation of XOR using $x_1\overline{x_2} + \overline{x_1}x_2$, the corresponding MLP is depicted in Fig. 3.4.

   - Note that there are 3 layers in this case also. The inputs are x_1 and x_2.
   - In the hidden layer there are two perceptrons. The top one outputs $x_1\overline{x_2}$; it is represented by the equivalent form $x_1 - x_2 > 0$. Similarly, the second perceptron in the layer outputs $\overline{x_1}x_2$; its equivalent representation is $x_2 - x_1 > 0$.
   - The output layer has a single perceptron which is an *Inclusive OR* gate; it has two inputs. So, its output is $x_1\overline{x_2} + \overline{x_1}x_2$ that is equivalent to the XOR function.

- Even though these two ANNs are very simple, early work on $MLPs$ exploited the results on these networks to highlight the fact it is possible to learn the weights connecting perceptrons (or neurons as they are called) in successive layers.

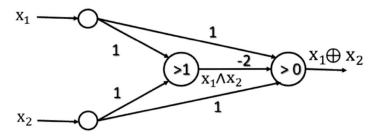

Fig. 3.3 Exclusive or represented as $x_1 + x_2 - 2x_1x_2$

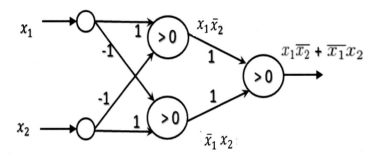

Fig. 3.4 Exclusive or represented as $x_1\bar{x_2} + \bar{x_1}x_2$

- *MLPs* are called *feedforward networks* as can be illustrated by using Fig. 3.4. The outputs of the neurons in layer i become the inputs to neurons in the $(i + 1)$th layer.
- If we observe Fig. 3.3, we see that the $(i + 1)$th layer gets inputs not only from the ith layer but also from the earlier layers. Typically, in an *MLP*, the weights connect neurons in two successive layers only.
- Learning in *MLP* networks amounts to starting with a set of initial weights and keep changing or updating the weights based on some criterion.

3.2.4 Training MLP Networks

The earliest and still the most popular algorithm for training *MLPs* is *backpropagation*. Before we consider the backpropagation algorithm, let us consider a related problem of training a single perceptron using the so called *delta rule*.

3.2.4.1 Delta Rule

Let us consider training a single perceptron.

- Let y^i_{obt} be the output obtained by the perceptron for input X_i.
- Let the target or expected output for the pattern X_i be y^i_{tar}.
- Let there be n training patterns given by $\{(X_1, y^1_{tar}), (X_2, y^2_{tar}), \ldots, (X_n, y^n_{tar})\}$.
- The idea is to start with some initial weight vector and update the weight vector so that *error between the target outputs and obtained outputs is minimized.*
- The error, $Error(W)$ is defined as

$$Error(W) = \frac{1}{2} \sum_{i=1}^{n} (y^i_{tar} - y^i_{obt})^2 \tag{3.2}$$

Here, $\frac{1}{2}$ is used for the convenience of calculus.
- We know that $y^i_{obt} = f_a(W^t X_i + b) = f_a(W^{Aug}.X^{aug}_i)$ where $W^{Aug} = (b, w_1, \ldots, w_d)^t$ and $X^{aug}_i = (1, x_{i1}, x_{i2}, \ldots, x_{id})^t$ are the augmented vectors that subsume the *bias* b into W.
- So, the process of learning W and b is converted into learning W^{aug}. Henceforth, we use W instead of W^{aug} for the sake of simplicity in notation. Correspondingly X^{aug}_i is called X_i.
- We assume that f_a is a linear function defined as $f_a(wsum) = wsum$. So, $y^i_{obt} = f_a(W^t X_i) = W^t X_i$ where W and X_i are augmented respectively.
- Finding the *optimal* W is done in the case of the delta rule by using gradient descent. The partial derivatives involved in computing the gradient of $Error(W)$ with respect to W are calculated by using the *chain rule*:

$$\frac{\delta Error(W)}{\delta w_j} = \frac{\delta Error(W)}{\delta y^i_{obt}} \cdot \frac{\delta y^i_{obt}}{\delta w_j} \tag{3.3}$$

- Note that

$$\frac{\delta Error(W)}{\delta y^i_{obt}} = -(y^i_{tar} - y^i_{obt}) \tag{3.4}$$

and

$$\frac{\delta y^i_{obt}}{\delta w_j} = x_{ij}, \quad for \ j = 0, 1, \ldots, d \tag{3.5}$$

by assuming that $w_0 = b$ and $x_{i0} = 1$ for $i = 1, 2, \ldots, n$.

- So,

$$\frac{\delta Error(W)}{\delta w_j} = \frac{\delta Error(W)}{\delta y^i_{obt}} \cdot \frac{\delta y^i_{obt}}{\delta w_j} = -(y^i_{tar} - y^i_{obt}).x_{ij} \tag{3.6}$$

- So, the gradient descent that employs the negation of the gradient will mean

$$W(k+1) = W(k) - \eta(-(y^i_{tar} - y^i_{obt})X_i) = W(k) + \eta(y^i_{tar} - y^i_{obt})X_i. \tag{3.7}$$

Table 3.2 Training data for the delta rule

Pattern	Value	y_{tar}
X_1	0	0
X_2	1	3
X_3	2	6
X_4	3	9

where the updated weight vector, $W(k+1)$ is obtained by updating the current weight vector, $W(k)$ and η is the learning parameter.

- It is called the *delta rule* or the *delta learning rule* because the difference (delta) between the target output and the obtained output is involved in the computation of the update.
- Note that the linear activation function is used instead of the linear threshold function, popularly used by Perceptron, to ensure that $\frac{\delta y_{obt}^i}{\delta w_j}$ can be computed; this is not possible when we use the linear threshold function.
- *Algorithm for Learning W*:

  1. Choose $k=1$ and initialize $W(k)$ with small values, and η with a small value.
  2. Consider each pattern X_i in the training set and update to get $W(k+1) = W(k) + \eta(y_{tar}^i - y_{obt}^i)X_i$. Set $k = k+1$. Update till all the patterns are considered; this is called an *epoch*.
  3. Stop if there is no change in the weight vector for an entire epoch, else iterate by going to step 2.

- *Example*:

  - Let us consider a function $g : \Re \to \Re$ given by $g(X) = 3X$. Let the training data be as shown in the Table 3.2.
  - We consider a simple perceptron with no bias that is shown in Fig. 3.5.
  - Let us initialize the value of weight $W(1)$ to 0.1 and η to 1.
  - The first pattern in Table 3.2 is input to the perceptron in Fig. 3.5, It is correctly classified as $X_1 = 0 \Rightarrow y_{obt}^1 = W.X_1 = 0 = y_{tar}^1$. So, $W(1)$ is not updated.
  - We consider $X_2 = 1$. The value of $y_{obt}^2 = W.X_2 = 0.2 \cdot 1 = 0.2$ and $y_{tar}^2 = 3$. So, $W(2) = W(1) + \eta(y_{tar}^2 - y_{obt}^2).X_2 = 0.2 + 1(3 - 0.2).1 = 3$.
  - Using the value of $W(2)$ all the patterns will be correctly classified. So, the algorithm stops and the weight value is 3.

- The example considered is very simple and it is primarily used to illustrate the *delta rule* based learning of the weight W. Also, it is a simple curve fitting/regression problem that may be viewed as a generalized version of the classification problem.

Fig. 3.5 Example
perceptron with bias, b = 0

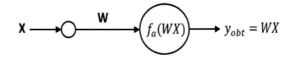

3.2.4.2 Backpropagation

Let us consider a multi-layer network (MLP).

- It will have $p(> 2)$ layers with the input layer (first layer), output layer (pth *layer*), and $p - 2$ hidden layers. For example, in Fig. 3.4 the value of p is 3 with one hidden layer.
- There will be n_q neurons in the qth layer for $q = 1, 2, \ldots, p$. In Fig. 3.4, there are 2 ($n_q = 2$) neurons in each of the input and hidden layers.
- Every neuron in layer q is connected to every neuron in layer $q + 1$ for $q = 1, \ldots, p - 1$. The connections in Fig. 3.4 exemplify this.
- The weight specified by w_{rs}^q is the weight associated with the connection between rth ($r = 1, 2, \ldots, n_q$) neuron in the qth layer and sth neuron in the $(q + 1)$th layer.

Some important properties of backpropagation may be summarized as follows:

- Let there be n training patterns given by $\{(X_i, y_t^{ip}), i = 1, 2, \ldots, n\}$, where y_t^{ip} stands for the target output at the pth layer (output layer) when X_i is input. Further, X_i is a vector of dimension n_1 and y_t^{ip} is a vector of dimension n_p.
- Let $wsum_{is}^{q+1}$ be the weighted sum input to the activation function at node s in the $(q + 1)$th layer for input X_i, where

$$wsum_{is}^{q+1} = \sum_{j=1}^{n_q} w_{js}^q y_o^{iqj}, \tag{3.8}$$

where y_o^{iqj} is the output obtained at the jth node of the qth layer for input X_i.
- We assume that the same activation function, f_a, is used at all the neurons in the entire MLP network.
- Note that w_{js}^q affects the final error through $wsum_s^{q+1}$.
- It is a feedforward neural network. So, when training vector X_i is presented at the input layer of the network, then y_o^{i1j} ($= X_{ij}$) is the output of the jth node in the first layer (input layer).
- The output of the jth node in layer q is given by y_o^{iqj} ($= f_a(wsum_{ij}^{q-1})$).
- Let the obtained output at the jth node in the output layer be y_o^{ipj}.
- We need to learn all the weights w_{rs}^q, $q = 1, \ldots, p$, $r = 0, 1, \ldots, n_q$, and $s = 1, \ldots, n_{q+1}$. Let W be the collection of these weights.
- Training the MLP is achieved by getting W that minimizes *squared error* across all the n patterns. It is given by

$$Error(W) = \sum_{i=1}^{n} \frac{1}{2} \sum_{j=1}^{n_p} (y_t^{ipj} - y_o^{ipj})^2. \tag{3.9}$$

- We use gradient descent and the update rule is given by

$$w_{rs}^q(k+1) = w_{rs}^q(k) - \eta \frac{\delta Error(W)}{\delta w_{rs}^q} \tag{3.10}$$

- We can compute $\frac{\delta Error(W)}{\delta w_{rs}^q}$, using the *chain rule* similar to the one used by the delta rule as

$$\frac{\delta Error(W)}{\delta w_{rs}^q} = \frac{\delta Error(W)}{\delta wsum_{is}^{q+1}} \cdot \frac{\delta wsum_{is}^{q+1}}{\delta w_{rs}^q} \tag{3.11}$$

- Note that $\frac{\delta wsum_{is}^{q+1}}{\delta w_{rs}^q} = y_o^{iqr}$.
- Let $error_{is}^{q+1} = \frac{\delta Error(W)}{\delta wsum_{is}^{q+1}}$. It is possible to view it as backpropagated error at node s in layer $q+1$ when X_i is input to the MLP.
- The process of backpropagation of error is initiated at the output layer; once we know $error_{ij}^p$ for node j in the output layer, we propagate it back to nodes in layer $p-1$, then to nodes in layer $p-2$, and so on till the nodes in the input layer.
- The error propagation is characterized by

$$error_{ij}^q = \sum_{s=1}^{n_q+1} error_{is}^{q+1} W_{js}^q f_a'(wsum_{ij}^q) \tag{3.12}$$

- The previous error computation is a result of the following chain rule:

$$error_{ij}^q = \sum_{s=1}^{n_q+1} error_{is}^{q+1} \cdot \frac{\delta wsum_{is}^{q+1}}{\delta y_o^{iqj}} \cdot \frac{\delta y_o^{iqj}}{\delta wsum_{ij}^p} \tag{3.13}$$

- In propagating the error back, we first start with the output layer (layer p)

$$error_{ij}^p = \frac{\delta Error(W)}{\delta wsum_{ij}^p} = \frac{\delta Error(W)}{\delta y_o^{ipj}} \cdot \frac{\delta y_o^{ipj}}{\delta wsum_{ij}^p} = (y_o^{ipj} - y_t^{ipj}).f_a'(wsum_{ij}^q) \tag{3.14}$$

- This is easy to compute because for input X_i we go through the forward pass to compute y_o^{iqj} for $j = 1, 2, \ldots, n_q$ and $q = 1, 2, \ldots, p$. Once we have y_o^{ipj}, we can compute $(y_o^{ipj} - y_t^{ipj})$.
- The quantity $f_a'(wsum_{ij}^q)$ can be computed because the form of f_a' is known in advance based on the functional form of f_a.
- Once we compute $error_{ij}^p$, for all the nodes in the pth layer (output layer), then we can propagate back, using the earlier equation, to get error of nodes in the previous

layer, and iteratively till we get error at every node in the MLP. We can use these errors to update the weights across the network for pattern X_i.

- The process is repeated for all the patterns; such an iteration over all the patterns is called an *epoch*. This process is repeated over several such epochs till some *termination criterion* on the error at the output layer is met.
- During the early days of MLP research, there was more effort on

  - Why linear threshold activation is inadequate? There was a need for an activation function f_a that is differentiable for the backpropagation of error. One of the most popular is the sigmoid function given by $f_a(x) = \frac{1}{1+e^{-x}}$.
  - How many hidden layers are required to learn a required function on an MLP? The universal function approximation theorem showed that one hidden layer is adequate to approximate any function. The radial basis function networks were based on this.
  - What happens if we use more hidden layers? The number of weights in the MLP network contribute to the dimensionality of the problem. So, more hidden layers mean more weights and a higher dimensional problem. With smaller training sets, the learnt MLP can overfit.

3.3   Convolutional Neural Networks

In the previous section, we have examined the MLP network. Some of the problems associated with it are:

1. The sigmoid activation function can have *vanishing or exploding gradient*; so, it is not the right activation function. This is also linked with how the initial weights are chosen.
2. Overfitting the training data can occur if the number of hidden layers/neurons is large; this happens if the training data is small.
3. Most of the backpropagation training scenarios used software simulations on slower machines; in the early days people were even restricting the weights to have integer values to run the simulations faster.

There are better and efficient processing platforms available now. We will consider the details associated with the activation functions and weight initialization in the next two subsections.

3.3.1   Activation Function

- *Earlier Activation Functions:*
  In the case of delta rule, we have seen the use of the linear activation function. It is not useful in dealing with any required nonlinearity across multiple layers

as it collapses multiple layers in the network into one; this is because the relation between the input and the final output will be through another linear function. This is similar to multiplication of several matrices giving rise to another matrix. This prompts the use of a nonlinear activation function. A popularly used nonlinear activation function is the *sigmoid function*. Some of its properties are:

- It is given by $f_s(x) = \frac{1}{1+e^{-x}}$. So, it maps any real number to a value in $[0, 1]$. Further, values of the input x above 5 will take the output close to 1 and values below -5 make the output close to 0.
- It needs to compute the exponential of the argument which could be time consuming.
- Its derivative is $f_s'(x) = f_s(x)(1 - f_s(x))$. As x tends to a large value, $(1 - f_s(x))$ tends to 0 (zero) and if it tends to a small value $f_s(x)$ tends to zero. So, in either case the derivative tends to 0. Thus the gradient can vanish.
- It is not zero centered. It assumes only positive values. This can affect the resulting output badly when there are many hidden layers in the MLP.
- If the *wsum* is small as the initial weights are small, then the derivative of the sigmoid function will be close to 0 and may even vanish. So, if there are more layers to be trained, then backpropagation may fail to update the weights in the earlier (closer to the input) layers as weights in such layers are considered for updation towards the end of error backpropagation. This is because of the *vanishing gradient* problem.
- On the contrary, if the weights are initialized with larger values, then it is possible to have the *exploding gradient problem* where the gradient can assume a value that is prohibitively large.
- A solution offered, to handle the zero-center problem, is in the form of the tanh function, f_t, that may be defined as

$$f_t(x) = 2f_s(2x) - 1. \tag{3.15}$$

- It is easy to see that f_t maps any real number to a value in the rang $[-1, +1]$.
- Both sigmoid and tanh functions are still used, even though they are not as popular as earlier as both may have difficulty with their gradients. It may lead to vanishing or exploding gradient problem which can impact the training accuracy and time.

- Activation Functions Popular with Deep Neural Networks (DNNs)

  - Rectified Linear Unit (ReLU): It is a popular activation function. It has the following characteristics:

    · It is defined as

    $$f_r(x) = \begin{cases} 0 & x \leq 0 \\ x & x > 0 \end{cases} \tag{3.16}$$

    · It is popular because it is computationally simpler.
    · It is used only in the hidden layers.

- It works well when x is positive. Its gradient vanishes when x is 0 or negative; so not useful for backpropagation when its input falls in this range. This is called the *Dying ReLU problem*.

  - Leaky ReLU: It offers a solution to the *dying ReLU problem*. Its properties are:

    - It is defined as $f_l(x) = max(0.01x, x)$. So, it is a variant of ReLU function.
    - It permits backpropagation for input values that are less than or equal to zero also. However, in this range predictions based on $f_l(x)$ may be inconsistent.
    - It trains faster then ReLU.

  - Softmax: It permits us to convert a vector of values to another vector of same size that has normalized values adding upto 1. Its characteristics are:

    - It is a mapping from $\Re^{n_p}$ to $(0, 1)^{n_p}$. It is specified as

$$f_{smax}(y_o^{ipj}) = \frac{e^{(y_o^{ipj})}}{\sum_{j=1}^{n_p} e^{(y_o^{ipj})}} \tag{3.17}$$

    - It is used at the output layer of a DNN to convert a vector of real numbers into vector of probabilities; the sum of the values of its outputs is 1.

3.3.2   Initialization of Weights

Different schemes have been used to initialize weights in the past.

- *Zero Weights*: Typically in perceptron training based on fixed increment rule, it is convenient to start with a zero weight vector and still guarantee convergence of the update algorithm when the classes are linearly separable. However, in the case of a DNN, *initializing all the weights to 0* or in general any constant value can lead to highly symmetric behaviour across the network leading every weight to be the same across the iterations.
- *Random weights*: It is possible to view a *deep neural network (DNN)* as a device that transforms the input, to match with the desired output, through successive layers. It is a lossy transformation. So, if we select the weights randomly, then the information loss in the initial layers may be so bad that the backpropagation algorithm may not be able to abstract the desired overall mapping even over a good number of iterations/epochs.
- *Smaller or larger weights:* In the previous subsection we have considered how initialization with smaller or larger weights can lead to vanishing or exploding gradient problems.

These issues associated with initialization were responsible for an appropriately normalized scheme to work. Some important normalization schemes that try to maintain zero mean and specified variance of the weights in the DNN are:

- *Xavier initialization and variants*:
  - Here the weights in layer q, $q = 1, \ldots, p$ are initialized by

$$w_{rs}^q \sim \left[-\frac{\sqrt{6}}{\sqrt{n_q + n_{q+1}}}, \frac{\sqrt{6}}{\sqrt{n_q + n_{q+1}}} \right] \tag{3.18}$$

    where weights are randomly drawn from a uniform distribution in the normalized range specified.
  - The bias for each neuron is initialized to 0 (zero).
  - This normalization is to ensure that the weights are chosen with a zero mean and a standard deviation that is a normalized version of 1.
  - It is a replacement of earlier normalization schemes that took into account only n_q.
  - This modification helps in ensuring that the activation outputs and gradients encountered in the backpropagation runs have variances that are neither too small nor too large.

- Kaiming Initialization:
  - A variant is proposed by Kaiming He et al. where values of weights w_{rs}^q are randomly chosen from the standard normal distribution and are multiplied by $\frac{\sqrt{2}}{\sqrt{n_q}}$ in this initialization.
  - This works better than Xavier initialization when ReLU activation is used.
  - It was observed that both training and testing errors converged to be requiring a smaller number of iterations/epochs to converge; 20 epochs appeared to be adequate in practice for a good performance.

- Typically these schemes conduct analysis based on looking at the variance of the product of weights and outputs of the neurons.

3.3.3 Deep Feedforward Neural Network

Before the year 2000, it was strongly believed that one or two hidden layers are adequate to deal with most of the machine learning tasks. One major observation was that backpropagation is based on gradient descent and it can only guarantee to reach a *locally optimal value* of the criterion function. This was the reason for *support vectors machines (SVMs)* to flourish, for more than two decades, as the machine learning benchmark tool as it guarantees globally optimal margin based learning in theory. However, in the past decade the earlier views ware significantly altered due to some important contributions in the area of deep learning. It became so important that every problem in the area of artificial intelligence (*AI*) is invariably solved using deep learning. Convolutional neural network is the popular feed forward deep learning architecture. Some of the contributions related to convolutional neural networks are discussed below:

- *Convolution:* It is a well-known operation in signal processing with applications to speech signals (one-dimensional) and images (two-dimensional).
- Let I be a two-dimensional image input which is represented as an array of size $r \times s$; So, I has r rows and s columns.
- Let the convolution template (or kernel) T be a smaller size pattern of size $M \times N$.
- The convolution output, O, that is an array of size $(r - M + 1) \times (s - N + 1)$ is given by

$$O(i, j) = f\left(\sum_{m=1}^{M} \sum_{n=1}^{N} I(i + m - 1, j + n - 1) T1(m, n)\right). \qquad (3.19)$$

The role of template T in convolution is to locate parts of the input image I that match with the pattern present in T.

- Function f may be defined as $f(x) = 1$ if $x > \theta$ else $f(x) = 0$ where θ is a threshold.
- Let us illustrate the convolution operation in two dimensions using the example in Fig. 3.6.

  - There are two parts labeled (a) (upper part) and (b) (lower part) in the figure. The input image in both the parts is the same. It is a 9×9 binary image array $INPUT$, of *character 7*, consisting of 81 pixels labeled $INPUT(1, 1)$ to $INPUT(9, 9)$. Note that $r = s = 9$ in this example.
  - It has a horizontal line segment in the top part against $INPUT(2, 1)$ to $INPUT(3, 9)$ (rows 2 and 3) and a vertical line segment in columns 7 to 9 across rows 2 to 9.

Fig. 3.6 Example convolution operation on input image of character 7

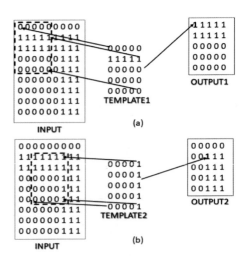

- In part (a), $Template1$ is used for convolution. It is a 5×5 binary pattern, $Template1$ of 25 binary pixels addressed by $Template1(1, 1)$ to $Template1$ $(5, 5)$. So, $M = N = 5$ here.
- Further, Note that in (a) $Template1$ is aligned with the top left part in $INPUT$ such that $Template1(1, 1)$ is aligned with $INPUT(1, 1)$ and $Template1$ $(5, 5)$ is aligned with $INPUT(5, 5)$.
- The pixel wise multiplication and addition as indicated in the equation for O gives us $OUTPUT1(1, 1) = f(5)$. If $\theta = 3$, then $f(5) = 1$. This value 1 is indicated in $OUTPUT(1, 1)$ in (a). If f is not used then we get the value $O(1, 1) = 5$.
- By shifting one position horizontally and multiplying and adding we get $OUTPUT1(1, 2)$. Further, while computing $OUTPUT1(1, 5)$, $TEMPLATE1$ will be aligned completely with the top 5 rows and rightmost 5 columns of $INPUT$. So, moving right further is not done which will make only a part of $TEMPLATE1$ align with a part of $INPUT$. Under these conditions, the other option is to create virtual columns filled with 0s. However, we consider alignment of the Nth column of the template with the sth column of the input and move no further.
- In order to compute $OUTPUT1(i, j)$ for $i = 2, \ldots, r - M + 1$ and $j = 1, \ldots, s - N + 1$ we need to align the top row of $TEMPLATE1$ with the ith row of $INPUT$.
- This results in the output image given by $OUTPUT1$ shown in (a). Note that this template has captured the horizontal lines in the input. It is popularly called as *mask* in image processing and *kernel/filter* is the popularly used term in $CNNs$.
- Similarly $TEMPLATE2$ in part (b) captures the vertical lines present in the character image in $INPUT$.
- This example is meant to illustrate the notion of convolution more than being a real mask for use in image convolution. Further, the threshold based function f is used here to get a binary output; such a function is not used in practice.

- *Feature Maps:* In a CNN, we will have multiple convolution layers. For example in Fig. 3.6 we have seen two different templates working on the same input image. $TEMPLATE1$ looks for horizontal lines in the input; this may be viewed as extracting one kind of feature. Similarly $TEMPLATE2$ looks for vertical lines; so extracts a different kind of feature. Each of the resulting outputs may be viewed as a *feature map*. In a more generic setting, we will have

  - Multiple templates/kernels each looking for a different kind of feature.
  - It is possible to have more than one occurrence of a feature in the same input image. For example, instead of character 7, if we consider the character 0 (zero) shown in Fig. 3.7 that has two horizontal (leftmost and rightmost) and two vertical (top and bottom) segments, then the same templates, $TEMPLATE1$ and $TEMPLATE2$ will each extract the respective features twice.
  - In practice, we may have images that are much larger in size compared to the small 9×9 input images shown in Figs. 3.6 and 3.7.

```
111111111
111111111
110000011
110000011
110000011
110000011
110000011
111111111
111111111
```

Fig. 3.7 Example of character zero with two horizontal and two vertical segments

- Also there can be a good number of templates each looking for one or more occurrences of the feature embedded in it. Correspondingly, there can be several feature maps one for each template.
- Each hidden layer may be viewed as made up of such multiple feature maps as many as the number of templates used in convolutions.
- The output of convolution is generally defined as

$$O(i, j) = f(\sum_{m=1}^{M} \sum_{n=1}^{N} w_{mn} I(i + m - 1, j + n - 1) T1(m, n)). \qquad (3.20)$$

In Fig. 3.6, the value of w_{mn}, $\forall m, n$ is taken to be 1. However, in a CNN these weights are learnt.

- An important aspect of learning these weights is that for each feature map we need to learn only $MN + 1$ weights where $M \times N$ is the size of the template and the extra 1 is to learn the bias term associated with the node in the hidden layer. This is an important characteristic of $CNNs$ and is called *weight sharing*.
- Further, the value of $MN + 1$ is much smaller in practice than the size of the image given by $r \times s$.
- We have assumed that the shifting of the template, after each multiply and add operations, is done by one column horizontally or one row vertically (by one pixel); in such a case the stride is 1. We can have strides of length 2 or more.

- Convolution and Pooling Layers: Each convolution layer has the input and hidden layers as shown in Fig. 3.8; a hidden layer has some L feature maps. So, the hidden layer will have $L \times M \times N$ neurons.

    - In a CNN there will be more than one such convolution layer. Typically after each convolution layer, there will be a pooling layer to reduce the dimensionality further.
    - A *pooling layer* is obtained from the features maps in the hidden layer of the previous convolution layer.
    - Let the size of each feature map be $u \times v$; so number of neurons in a feature map is uv.
    - Let the pooling be done by using a window of size $k \times k$, where k divides both u and v, over the feature map. This is done by considering $k \times k$ neurons in

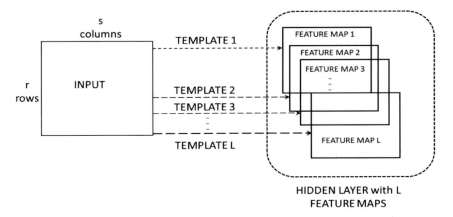

Fig. 3.8 Convolution layer with multiple feature maps in the hidden layer

the feature map at a time; the window is moved horizontally and vertically in a non-overlapping manner.

– In each window region of k^2 neurons, the respective k^2 outputs are pooled to output one value that is stored in the corresponding location output of the pooling layer.

– The output of the pooling layer is given by

$$Poolout(i, j) = g(\{fmo((i-1)k+1, (j-1)k+1), \ldots, fmo(ik, jk)\})$$

$$(3.21)$$

where $fmo(p, q)$ feature map output of the neuron in the pth row and qth column of the feature map. Observe that $i = 1, \ldots, \frac{u}{k}$ and $j = 1, \ldots, \frac{v}{k}$.

– Note that the argument of g is a set of k^2 elements across rows $(i-1)k$ to ik (k rors) and columns $(j-1)k$ to jk (k columns). They are the outputs of neurons in the chosen $k \times k$ region in the feature map.

– The function g itself could popularly be the *max, average*, or $L_2 - norm$ of the k^2 values in the set.

• The overall architecture of the CNN will consist of several convolution layers; after each convolution layer there will be a pooling layer with the output of the feature maps in the layer forming the input of the pooling layer. The output of the pooling layer will be the input of the next convolution layer.

• Typically the final output layer of the CNN will be a fully connected layer that is connected to all the neurons in the previous layer.

• The CNN is trained using backpropagation. The error is propagated back from a layer to the previous layer through the relevant weights.

Some important properties of CNN are that

• It is the *state-of-the-art tool* for classification and prediction.

- It has been successfully used in large-scale applications where both the number of training patterns and/or the dimensionality of the data is large. In fact it works well only when the training data is large.
- It became popular because of its applications in image processing and speech processing applications.
- One of the important outcomes is a variant that has become popular in network applications in the form of *graph convolutional net* (*GCN*).

3.4 Recurrent Networks

Earlier in this chapter, we discussed MLP and CNN models. Some of the limitations associated with them are:

- They expect inputs of predetermined size and transform them into fixed-size vectors. In contrast, many real-world problems have an unknown size, such as machine translation, document classification tasks, which makes MLP and CNN type networks unsuitable for these applications.
- In many applications such as sentiment classification, sentence classification, etc., the input is a sequence of words and the computation at a step of the sequence depends on the current word and the previous words too. But MLP assumes that all the inputs are not dependent on each other and thus cannot process these inputs. Therefore, we require tools to deal with sequence data, where previous words also effect the computations at a later step.

Recurrent Neural Networks (RNNs), Long Short Term Memory ($LSTM$), Gated Recurrent Units (GRUs) are developed to solve the problems mentioned above. In the next two subsections, we discuss the RNN and the $LSTM$ models in detail.

3.4.1 Recurrent Neural Networks

A *Recurrent Neural Network* (*RNN*) is a multi-layered model that processes inputs sequentially. Some important characteristics are:

- RNN is a neural network model where previous outputs play a major role in determining the next output. These models have shown great success in many sequential tasks, especially in the *natural language processing* (*NLP*) domains.
- For example, a character level RNN considers each word as a single input (sequence), each character in the word as an element of the sequence, and each successive element is called a time step.
- RNNs use the same set of parameters for all the time steps of an input, which not only avoids overfitting but also learns dependencies between the elements at different time steps of the input. So, RNN performs the same operation on each element of the serial input. Thus these models are called recurrent.

- On the contrary, in a *vanilla neural network*, each input element is associated with a different set of weights due to which the network cannot work with serial inputs of varying sizes.
- *RNN* has *hidden states or units* which encapsulate the relations between elements of a serial input. Hidden states are also interpreted as *memory units*.
- Output calculation at each time step depends on the information present in the hidden state and the current input, thereby updating the weights of the model and the information in the hidden state.
- Figure 3.9 shows one time step of *RNN*. Figure 3.10 depicts the complete architecture of *RNN*.
- At each time step t, it takes the one-hot encoding of an element from the input sequence and outputs a vector whose each entry denotes the probability of the corresponding element being the next element in the sequence.

3.4.1.1 Working of Recurrent Neural Networks

Let us denote the input, hidden and output states at step t by x_t, hs_t and hy_t respectively. Initial hidden state hs_0 is generally initialized by zeros and x_1 (initial input) is a one-hot vector of the first element in a serial input.

- Current state of RNN is calculated by:

$$hs_t = \sigma \left(W_s hs_{t-1} + W_x x_t \right) \tag{3.22}$$

where x_t is the present input, hs_{t-1} is the hidden state at time step t-1 and hs_t is the new hidden state at time step t. σ is an activation function (tanh or ReLU). W_s and W_x are the collection of trainable weight parameters.
- Output state at time step t is calculated using the current hidden state

$$hy_t = W_y hs_t \tag{3.23}$$

where hy_t and hs_t are the output vector and the hidden state at time step t and W_y are the weights.
- To convert the output to a probability distribution (which is required for many tasks such as classification), softmax activation is used on hy_t,

$$o_t = \text{softmax}(hy_t) \tag{3.24}$$

- *RNN*s can have output at each time step or at only the final step. For example, task of classifying the entire sequence generates only one output at the last time step with no outputs at the intermediate time steps.

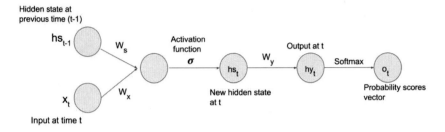

Fig. 3.9 A single layer of *RNN*

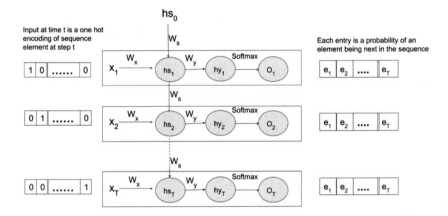

Fig. 3.10 Complete architecture of *RNN*

3.4.1.2   Backpropagation Through Time

RNN is also trained using the backpropagation algorithm. The recurrent network is a time sequence model. Therefore, backpropagation means going back in time and hence is called Backpropagation through time (*BPTT*).

- First of all, the total error for an input will be the sum of errors from each step.

  – Suppose loss at each step(t) is calculated by cross entropy between predicted vector($\hat{o}_t$) and the actual one hot encoding(l_t) of the correct output word. The total error for all the time steps is calculated as follows:

$$J(\hat{o}, o) = -\sum_{t=1}^{T} \hat{o}_t log(l_t) \qquad (3.25)$$

$J(\hat{o}, o)$ is the total error and T is the total number of time steps.

- Similar to errors, gradients are also summed up over all the time steps. We get the following equations corresponding to W_s, W_h and W_y:

$$\frac{\partial J}{\partial W_s} = \sum_{t=1}^{T} \frac{\partial J_t}{\partial W_s} \tag{3.26}$$

$$\frac{\partial J}{\partial W_h} = \sum_{t=1}^{T} \frac{\partial J_t}{\partial W_h} \tag{3.27}$$

$$\frac{\partial J}{\partial W_y} = \sum_{t=1}^{T} \frac{\partial J_t}{\partial W_y} \tag{3.28}$$

Here J_t is the error at time t. Each term in the summation will be evaluated similarly. So we will focus on the error term at time step t.

1. First we compute gradients wrt parameter W_y. The derivation of J_t wrt W_y depends only on the current time t. Formally, J_t depends on the predicted label ($\hat{o}$) (3.25) which depends on hy_t (3.24) and hy_t is a function of W_y (3.23). Thus, using the chain rule of differentiation we get the following equation:

$$\frac{\partial J_t}{\partial W_y} = \frac{\partial J_t}{\partial W_y} = \frac{\partial J_t}{\partial \hat{o}_t} \frac{\partial o_t}{\partial hy_t} \frac{\partial hy_t}{\partial W_y} \tag{3.29}$$

2. The process of calculating gradients wrt to W_x and W_s is different. Here we will calculate the gradients wrt W_s, and the same process can be repeated for the other.

   - Following the same steps as used for calculating the gradients for W_y and further noting that in RNN, W_s parameters are shared at all the time steps because of which changes in W_s will effect the error at time step t (J_t) even when $hs_1, hs_2, \ldots, hs_{t-1}, hs_t$ states are being computed. We get the following equation using the points mentioned above.

$$\frac{\partial J_t}{\partial W_s} = \sum_{q=0}^{t} \frac{\partial J_t}{\partial \hat{o}_t} \frac{\partial \hat{o}_t}{\partial hs_t} \frac{\partial hs_t}{\partial hs_q} \frac{\partial hs_q}{\partial W_s} \tag{3.30}$$

   - More generally, the third term in Eq. 3.30 is a chain of derivatives. As hs_k is a function of hs_{k-1} which depends on hs_{k-2} and this continues with first hidden state depending only on W_s. Based on this, the above equation can be rewritten as:

$$\frac{\partial J_t}{\partial W_s} = \sum_{q=0}^{t} \frac{\partial J_t}{\partial \hat{o}_t} \frac{\partial \hat{o}_t}{\partial hs_t} \left(\prod_{p=q+1}^{t} \frac{\partial hs_p}{\partial hs_{p-1}} \right) \frac{\partial hs_q}{\partial W_s} \tag{3.31}$$

- For example, while calculating gradient at time step $t = 3$ and examining the effect of the change in W_s on J_3 when hs_1 is being evaluated.

$$\frac{\partial J_t}{\partial W_s} = \frac{\partial J_3}{\partial \hat{o}_3} \frac{\partial \hat{o}_3}{\partial hs_3} \frac{\partial hs_3}{\partial hs_2} \frac{\partial hs_2}{\partial hs_1} \frac{\partial hs_1}{\partial W_s} \tag{3.32}$$

3.4.1.3   Vanishing and Exploding Gradients

The problems of vanishing and exploding gradients occur in deep feedforward neural networks and are already discussed. These problems also exist in $RNNs$. In this subsection, we discuss these problems and some existing solutions.

1. Recurrent Neural Networks suffer from the short-term memory problem, i.e., RNNs cannot learn dependencies between far apart elements. The diminished information from previous time steps is the consequence of the vanishing gradient problem.
2. Therefore, Vanilla $RNNs$ face problems dealing with long-range dependencies. For example, in the sentence, "Tyson had a trip to a hill station with his friend", "his" is used for "Tyson", and to figure this relation, $RNNs$ will have to remember a lot of information.
3. More formally, the expanded Eq. 3.30 includes the chain of derivatives of $\frac{\partial hs_3}{\partial hs_k}$ that depends on the derivative of the activation functions. The value of the derivatives of tanh or sigmoid activation functions can reach 1 or 1/4, respectively.
4. Also, gradients of tanh and sigmoid become 0 during saturation. As a consequence, the gradients of neurons from far away steps approach 0. The multiplication of such small values significantly shrinks the gradient, and after a few steps, it vanishes, and hence those neurons will not learn anything.
5. Some existing solutions for the Vanishing Gradient problem are:

   - Proper initialization of the W matrix needs to be used.
   - A better solution is to use a variant of RNN, such as Long Short-Term Memory ($LSTM$) or Gated Recurrent Unit (GRU). Both these models can overcome the problem of vanishing gradients.

6. Another problem with RNN is the exploding gradient, the opposite of the vanishing gradient, in which gradients become very large (will have values as NaN). This can be solved by using a threshold value as a cap on all the gradients.

3.4.2 Long Short Term Memory

The short term memory of Recurrent Neural Networks makes difficult for RNNs to carry information from previous time steps that are far apart because gradients become very small, and no learning can take place from that point. Long Short Term Memory ($LSTMs$), an improvement over $RNNs$, were developed to solve this vanishing gradient problem and handle long range dependencies between the elements. Some essential characteristics of $LSTMs$ are:

1. Major difference between RNN and $LSTM$ is their cell structure. Each RNN module is a simple single layer neural network structure, while each $LSTM$ module is a more complicated structure and uses four gates or four neural network layers.
2. LSTM core idea is its cell state and its gates (input, forget, output).

   • Cell state (represented as C_t) at time t carries and passes only the appropriate information during training.
   • Gates are used to distinguishing the important and the irrelevant information from the cell state and based on the importance score update the cell state.
   • These gates are composed of multiplication operation and a neural network with a sigmoid layer.

3. Just like humans tend to forget unimportant words and remember only the main parts of a speech, gates in $LSTM$ also help learn only the relevant information. Hence, they solve problems associated with the short term memory of RNN.
4. Similar to $RNNs$, $LSTMs$ too have hidden state hs_t at each time t.

3.4.2.1 Different Gates Used by $LSTM$

$LSTM$ uses various GATES for different purposes. All these gates are neural networks.

• Sigmoid layer is used in almost all the gates to determine the information to be updated and the information to be discarded.
• Sigmoid function outputs values between 0 and 1, with 1 representing the most important information and 0 representing the least important information.
• If any value in the cell is multiplied by 0, then the cell forgets that information and does not let that information pass through; otherwise, the value is fed to the later time steps.

Now we discuss important units of $LSTM$ architecture. In the text and equations below, we use t to denote the current time step, hs_{t-1} to represent the hidden state at the previous time step t-1, and x_t as input at time t.

• *Forget Gate* decides which values to be discarded from the cell state at the previous time step.

- It uses a sigmoid layer that takes the previous hidden state hs_{t-1} along with the current input x_t and gives values between 0 and 1.
- All important information will have values closer to 1. The forget gate can be described as follows:

$$fg_t = \sigma(W_{fg}[hs_{t-1}, x_t] + \beta_{fg}) \tag{3.33}$$

Here W_{fg}, β_{fg} are the weights and bias terms associated with the forget gate layer. σ is an activation function. fg_t is the output of the forget gate at time step t.

- The next step is to determine the information to be included in the cell state.

  - It is done by a group of 2 layers.

    · *Input gate layer* uses a sigmoid layer that takes the previous hidden state and the current input and decides which information to update. Equation 3.34 describes this step.

$$gi_t = \sigma(W_{gi}[hs_{t-1}, x_t] + \beta_{gi}) \tag{3.34}$$

Here W_{gi}, β_{gi} are the parameters of the input gate layer.

    · *tanh layer* outputs values between -1 and 1. It gives a new set of entries $\hat{ci}_t$ that can be included in the cell state. Equation 3.35 describes this step.

$$\hat{ci}_t = tanh(W_{ci}[hs_{t-1}, x_t] + \beta_{ci}) \tag{3.35}$$

Here W_{ci}, β_{ci} are the parameters of tanh layer.

  - Multiplication of these two outputs determines the useful entries of $\hat{ci}_t$ with the help of sigmoid output gi_t.

$$z_t = gi_t * ci_t \tag{3.36}$$

- The next step is to form the new cell state C_t from the information calculated so far. Remember that fg_t knows what to throw away and what to keep for further states.

  - To form a new cell state, the first step is to multiply the cell state C_{t-1} with the forget vector fg_t.
  - This helps the cell to forget unimportant information by multiplying with a value closer to zero, which is determined by the fg_t entries. That way, it can focus only on appropriate part of the sequence until the previous time step.
  - We then add z_t (the input gate output) to the resulting product. The result is the new cell state, which contains the updated, appropriate information. The following equation describes these steps.

$$C_t = fg_t * C_{t-1} + z_t \tag{3.37}$$

where z_t is defined in Eq. 3.31.

- The next important gate layer is the *output gate layer*, which determines the next hidden state based on the new cell state just formed.

  - A sigmoid layer is used to decide the information from C_t to pass on to the next states. It takes x_t and C_{t-1} as inputs.
  - The next step is to use a tanh layer on the new cell state C_t.
  - Finally, multiplication of these sigmoid and tanh outputs will determine the information for the next hidden state. These steps can be described by the following equation:

$$og_t = \sigma(W_{og}[hs_{t-1}, x_t] + \beta_{og})$$
$$hs_t = og_t * tanh(C_t)$$

(3.38)

where W_{og}, β_{og} are the parameters of the output gate layer. hs_t is the new hidden state at time step t.

- Figure 3.11 shows a complete layer of the *LSTM* model.

  - This figure shows all the steps that we explained above to generate a new hidden and a cell state by using previous hidden state, current cell state, current input and all the gates.
  - In this figure, σ represents the sigmoid layer. Each blue circle denotes one of the layers described in the text above, and each red circle represents a mathematical operation (element-wise multiplication or addition).

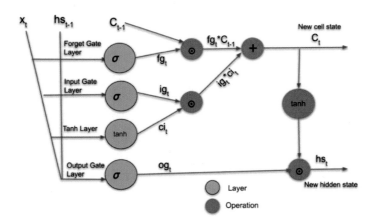

Fig. 3.11 LSTM complete architecture

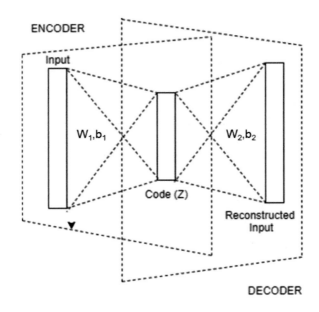

Fig. 3.12   Autoencoder architecture

3.5   Learning Representations Using Autoencoders

An autoencoder is a popular unsupervised model for learning representations in a low dimensional space.

- It is a neural network that employs a non-linear transformation on the input to compress it. This is done so that the original data can be reconstructed using this low-dimensional representation.
- An autoencoder incorporates an encoder and a decoder. The encoder compresses the input. The decoder decompresses the compressed input to get back the original input. Another important component is the code, also known as the bottleneck, which is the compressed representation of the input.
- An ideal autoencoder should be sensitive to the input to learn a less lossy reconstruction but, at the same time, should not learn an identity mapping.
- Significant applications of autoencoders are dimensionality reduction and representation learning. The recent development of variational autoencoders makes autoencoders useful as generative models also.
- Figure 3.12 shows the architecture of a simple autoencoder.

    - As shown by the outer boxes in the figure, the encoder comprises the input and the hidden layers, while the decoder is made up of the hidden and the output layers.
    - Input, output, and hidden layers can have any number of units (neurons).

- W_1, b_1 and W_2, b_2 indicate the weights, bias between input and hidden layers and hidden and output layers respectively.

  · In some cases, these weights can be tied together, such that $W_2 = W_1^T$, which is sometimes used to avoid overfitting as the number of trainable parameters is less in this setting.

- An autoencoder computes the compressed representation of the input as follows:

  - The Encoder receives the input (x) (input layer) and computes the latent representation (code) as $h = \sigma(W_1 x + b_1)$. This is fed to the decoder which outputs the reconstructed input (z) as $z = \sigma(W_2 x + b_2)$, where σ is an activation function.
  - Autoencoders are trained using gradient descent, and parameters are learned using the backpropagation rule as used by $MLPs$.
  - Loss function of autoencoder depends on the input x and the output z as the loss should reflect the deviation of the reconstructed input (output of autoencoder) from the input.

    · One possible loss function is L2 Norm. Let the dataset contains $x_1, x_2, \ldots, x_n$ samples, where n is the number of samples in the dataset. The L2 norm loss is given by:

$$Loss = \frac{1}{2} \sum_{i=1}^{n} |x^i - z^i|^2 \tag{3.39}$$

    · It is clear from the loss function that there is no role of the label information during training; thus, it is an unsupervised learning scheme. But autoencoders can be trained in a supervised manner for a specific downstream task such as the classification task.
    · For the supervised classification task, some fully connected layers with the last layer being the softmax layer, are appended. The model is trained in an end to end fashion using cross entropy loss, which leverages the label information. In this case, the information loss would be less.

3.5.1 Types of Autoencoders

There are many variations of autoencoders. In this subsection we briefly describe some of them.

- *Sequence-to-Sequence Autoencoder:* It uses recurrent neural networks for encoder and decoder operations. These autoencoders first convert the entire sequence to a single lower dimension encoding, following which the decoder tries to get back the sequence from this encoding.
- *Deep Autoencoder:* This is an extension of vanilla autoencoder that has many layers in the encoder and the decoder part. The first set of layers compresses the

input while the next set of layers (decoder) will reconstruct the input from the latent representations. Also, as we go deeper, more high order or more abstract features are learned.

- *Undercomplete Autoencoder*: This variant develops a generalized model with the encoder's output dimension (code dimension) smaller than the input dimension. These autoencoders can ensure that the model is not copying the input and is learning important data distribution features because a smaller code dimension restricts the information flowing through the model. This type of model only constraints the number of hidden units in the bottleneck layer. And there can be cases where the hidden layer has only one neuron while encoder and decoder having abundant capacity tend to overfit the data and, therefore, couldn't learn anything meaningful.

- *Regularized Autoencoders*: These autoencoders provide the ability to learn other properties of data instead of copying the input to the output even if the encoder and decoder have the superabundant capacity or the encoder output dimension is equal to or greater than the input dimension. Regularized autoencoders leverage a loss function which helps learn only the variations and not the redundancies in the data and further avoids overfitting.

- Various regularization techniques are used in order to prevent encoder and decoder from learning the identity functions.

    - *Denoising Autoencoders*: These autoencoders add some random gaussian noise to the inputs before training. However, the model still reconstructs the uncorrupted data because the model loss depends on the original input and not on the noisy input. This acts as a regularizer and helps autoencoders distinguish more essential parts of the input as these autoencoders try to undo the corruption. The loss function is as follows:

$$Loss = \frac{1}{2} \sum_{i=1}^{n} |x^i - \hat{z}^i|^2 \tag{3.40}$$

    Here $\hat{z}^i$ is the output of the model with input being the corrupted data. Same mechanism can be applied at any layer of the autoencoder.

    - *Sparse Autoencoder*: This variant minimizes the number of non-zero entries in the latent representation. It constrains the capacity of the model by penalizing the activations within the hidden layer, and hence without any limitation on the number of nodes in the hidden layer, the model can learn the input data distribution irrespective of the encoder and decoder capacity.

3.6 Summary

Deep learning is an important topic that has found applications in several areas.

- The availability of large scale datasets and powerful computing platforms have played an important role making deep learning possible.
- *Deep neural networks* form the *de facto* tools for deep learning.
- Perceptron is one of the earliest and the most basic neural network models. It forms the basis for a variety of neural network models.
- The need and importance of MLPs is considered next. Backpropagation is the training algorithm that was important in training MLP networks.
- The difficulty in increasing the number of layers was analysed to identify the vanishing and exploding gradient problems.
- Important contributions behind the design and training of deep neural networks in the form of activation functions like ReLU and softmax are examined.
- Another important contribution behind the success of deep neural networks is the weight initialization and updating.
- Other factors that impacted deep learning include convolution, pooling and weight sharing.
- Several important deep learning models including CNN, RNN, $LSTM$ and autoencoders are considered.
- CNNs have been extensively used in image processing and speech processing.
- For analysing sequence data RNNs and $LSTM$ are popularly used. They find applications in natural language processing and biological sequence data.
- Autoencoders are the most popular dimensionality reduction tools that can compress input data using a non-linear transformation.
- We considered some of the important properties associated with CNNs, RNNs and autoencoders, the difficulties in training these models, and solutions provided.
- The deep learning models are important in the context of network data analysis. We will consider specific roles of CNNs, autoencoders and RNNs in the context of network embeddings in the later chapters.

Bibliography

1. Bishop CM (2005) Neural networks for pattern recognition. Oxford University Press
2. Murty MN, Raghava R (2016) Support vector machines and perceptrons. Springer briefs in computer science
3. Loiseau JCB (2019) Rosenblatt's perceptron, the first modern neural network, https://towardsdatascience.com/rosenblatts-perceptron-the-very-first-neural-network-37a3ec09038a/
4. Mazur M (2015) https://mattmazur.com/2015/03/17/a-step-by-step-backpropagation-example/
5. Nielsen MA (2015) Neural networks and deep learning, vol 2018. Determination Press, San Francisco, CA, USA

6. Mhaskar HN, Micchelli CA (1994) How to choose an activation function. In: Advances in neural information processing systems, pp 319–326
7. Wu J (2017) Convolutional neural networks. Published online at https://cs.nju.edu.cn/wujx/teaching/15CNN.pdf
8. Britz D (2015) Recurrent neural networks tutorial, part 3 - backpropagation through time and vanishing gradients, http://www.wildml.com/2015/10/recurrent-neural-networks-tutorial-part-3-backpropagation-through-time-and-vanishing-gradients/
9. Wolf W (2016) Recurrent neural network gradients, and lessons learned therein, http://willwolf.io/2016/10/18/recurrent-neural-network-gradients-and-lessons-learned-therein/
10. Gupta DS (2017) Fundamentals of deep learning - introduction to recurrent neural networks, https://www.analyticsvidhya.com/blog/2017/12/introduction-to-recurrent-neural-networks/
11. Olah C (2015) Understanding LSTM networks, https://colah.github.io/posts/2015-08-Understanding-LSTMs/
12. Srivastava P (2017) Essentials of deep learning : introduction to long short term memory, https://www.analyticsvidhya.com/blog/2017/12/fundamentals-of-deep-learning-introduction-to-lstm/
13. Nguyen M (2018) Illustrated guide to LSTM's and GRU's: a step by step explanation, https://towardsdatascience.com/illustrated-guide-to-lstms-and-gru-s-a-step-by-step-explanation-44e9eb85bf21/
14. Goodfellow I, Bengio Y, Courville A (2016) Deep learning. MIT Press, http://www.deeplearningbook.org/
15. Jordan J (2018) Introduction to autoencoders, https://www.jeremyjordan.me/autoencoders/
16. Choi HI (2019) Lecture 16: autoencoders (Draft: version 0.7. 2)

Chapter 4
Node Representations

4.1 Introduction

A major issue in implementing ML schemes is the dimensionality of the data. If there is a sufficient amount of training data, then it is easy to train the ML models successfully, especially by using deep learning (DL) methods as they can learn the appropriate representations. However, even these DL models can succeed only if there are powerful machines and large datasets for learning the model.

In most of the current day applications, there is still a challenge to deal with high-dimensional datasets where the number of data points is not too large. For example, in social network analysis, we have to deal with adjacency matrices of size $N \times N$, where N is the number of nodes in the network represented as a graph. Here the dimensionality of each node is as large as the number of nodes in the network.

There are several applications involving networks. For instance, discovering new patterns in a drug-disease network helps develop new treatments, recommending friends in a social network, clustering publications in a citation network to find related domain papers, and find communities of users in a user interest network, which helps in appropriate broadcasting of news.

So, there is a natural requirement to employ ML tasks on the network datasets to carry out meaningful analysis of the underlying data. The downstream ML tasks include classification and clustering. In such cases, we require a low-dimensional representation of the network dataset. In this chapter, we deal with embedding network data in a low-dimensional space. Specifically, we examine various embedding schemes that can be used to embed nodes and edges in a low-dimensional space to facilitate the required downstream ML tasks.

The node-level downstream tasks include node classification, node clustering, recommendation, link prediction, and visualization. In this chapter, we discuss some node representation techniques based on random walk, matrix factorization, or deep learning methods. Further, some algorithms learn representations in an unsupervised

M. Aggarwal and M. N. Murty, *Machine Learning in Social Networks*,
SpringerBriefs in Computational Intelligence,
https://doi.org/10.1007/978-981-33-4022-0_4

setting while others learn in a supervised setting. We finally present comparisons of these algorithms according to their performance on downstream tasks.

4.2 Random Walk Based Approaches

To learn embeddings, the algorithm should at least consider the network structure. The most straightforward representation can be formed using the adjacency vector that only captures first-order neighborhood structure but has many disadvantages, including high dimensionality and sparsity. In this section, we discuss random-walk based node representation algorithms.

A random walk rooted at node v_i is a sequence of nodes $\{v_i^1, v_i^2, \ldots, v_i^l\}$ where, v_i^k is the vertex at step k in the walk and is randomly selected from the neighbors of the node at the previous step in the random walk i.e., from neighbors of v_i^{k-1}. Random walk based algorithms learn node embeddings by transforming the network into a collection of random walks, treating them as sentences in a language and applying the skip-gram model.

4.2.1 DeepWalk: Online Learning of Social Representations

DeepWalk is an unsupervised method based on the random walk technique. It learns representations of nodes by modeling a set of random walks. Let us consider a graph $G(V, E)$ where V is the set of N nodes, and E is the set of edges.

- DeepWalk aims to learn the latent embeddings of the nodes. For this, DeepWalk introduces a mapping function $g : v_i \in V \mapsto R^{N \times d}$, where $g(v_i) \in \mathcal{R}^d$ is the low-dimensional representation of node v_i using dimension d.
- A set of ρ walks of length L is sampled for each node in the graph. At each step of a walk, a node is sampled uniformly from the previous node neighbors in the walk, which is repeated until the maximum length L is achieved.
- For example, consider the graph in Fig. 4.1, some random walks rooted at vertex g can be: (g, d, e, f, a, b, c) (orange color arrows depict this walk), (g, d, e, f, c, b, a) and (g, d, c, b, a, f, e).

Fig. 4.1 An illustration of random walk rooted at node g

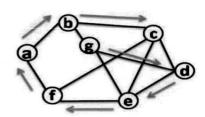

- Following the walk generation, it exploits the skip-gram model by considering each sampled walk as a sentence and the nodes of the walk as the words in the sentence to learn the latent representations of the nodes.

  - Like the context of a word, the context of a node in a walk consists of the nodes appearing on the right side and the left side within a window of size k around the given node.
  - Let us consider in Fig. 4.1 the random walk g, d, e, f, a, b, c with starting vertex g. In this example, if we consider a window of size $K = 2$, then the context of node f is the collection of nodes $\{d, e, a, b\}$.

  More precisely, for each walk in the set, the following steps are performed.

- DeepWalk iterates over each node v_i, called the source node, in the walk, and first maps the node v_i to the respective current representation $g(v_i)$ and then traverses through all its context nodes.
- Finally, DeepWalk maximizes the likelihood of observing the neighboring nodes (i.e., the context nodes) given the representation of the source node v_i, described by the following optimization problem:

$$\min_{g} - log \ Pr(v_{i-k}, \ldots, v_{i-1}, v_{i+1}, \ldots, v_{i+k} | g(v_i)) \qquad (4.1)$$

Here $\{v_{i-k}, \ldots, v_{i-1}, v_{i+1}, \ldots, v_{i+k}\}$ is the set of context nodes of v_i using a window size k. The minimization is done over g.

- Further, some relaxations are made to reduce the computational cost.

  - First, DeepWalk removes the ordering constraint and maximizes the probability of any context node without using the information of its distance from the source node.
  - Second, DeepWalk exploits the conditional independence assumption, i.e., given the representation of v_i, the probability of observing a context node is independent of other context nodes.

- Hence, Eq. 4.1 can be approximated using these assumptions as following:

$$\min_{g} - \sum_{j=i-k, j \neq i}^{j=i+k} log \, Pr(v_j | g(v_i)) \qquad (4.2)$$

Here $Pr(v_j | g(v_i))$ is the probability distribution which can be modeled using logistic regression.

- But this would lead to N number of labels (N is the number of nodes), which could be very large. Thus, to approximate the distribution $Pr(v_j | g(v_i))$, DeepWalk uses *Hierarchical Softmax*. Each node is allotted to a leaf node of a binary tree, and the prediction problem becomes maximizing the likelihood of a particular path in the tree.

The architecture of DeepWalk learns embeddings such that nodes will be closer in the embedding space if they share the same neighborhoods in the graph and hence preserving second and high order proximities.

4.2.2 Scalable Feature Learning for Networks: Node2vec

node2vec is another random walk based algorithm to learn node representations. Let $g : V \mapsto \mathcal{R}^{N \times F}$ be the function that maps nodes to their F dimensional embedding vectors, which node2vec aims to learn with $g(i)$ representing the embedding of node i. We first discuss the random walk generation procedure of node2vec.

Breadth-First Search (BFS) strategy limits the search to nearby nodes, i.e., local neighborhood, and captures a microscopic view that is essential and sufficient to infer the structural equivalences. Hence, BFS samples neighborhoods that lead to node representations such that structurally equivalent nodes remain closer in the embedding space. On the other hand, Depth First Search (DFS) can move far away from the root and represent the macroscopic view, which characterizes the communities according to homophily.

Building on this, node2vec designs a flexible biased random walk (neighborhood sampling strategy) that interpolates between breadth-first and depth-first strategies to traverse (sample) neighborhoods:

- Suppose w_i is the ith node in a random walk rooted at node u. If $(u, s) \in E$, node w_i is produced by distribution $P(w_i = s | w_{i-1} = u) = \frac{\pi_{us}}{Z}$, else it is 0. Here, π_{us} is the transition probability from u to s and Z is a normalizing constant.
- Formally, node2vec defines a second order walk, i.e., if the walk just crossed the edge (z, u), it selects the next node in the walk from the neighbors of u. Let that node be s, which is at distance d_{zs} of either 0 or 1 or 2 from node z. Thus, node2vec assesses the transition probability π_{us} for traversing edge (u, s) to visit node s.
- node2vec fixes the normalized transition probability π_{us} as:
  - if $d_{zs} = 1$, then $\pi_{us} = 1.w_{us}$
  - if $d_{zs} = 0$, then $\pi_{us} = \frac{1}{p}.w_{us}$
  - if $d_{zs} = 2$, then $\pi_{us} = \frac{1}{q}.w_{us}$

  Here w_{us} is the weight of edge (u, s) and the probability is managed by p and q parameters (discussed below).
- For instance, consider the graph in Fig. 4.2. We assume that we have reached node g from node e after traversing the edge (e, g). The walk visits the next node, a neighbor of node g, i.e., one of $\{b, c, f, e, d\}$. Note that the set of distances between node e and $\{b, c, f, e, d\}$ is $\{2, 1, 1, 0, 2\}$. Different colors show nodes with different distances from node e. Each edge label shows the transition probability according to the distance of the respective node from node e.
- Two parameters control the transition probability, return parameter p, and in-out parameter q.

Fig. 4.2 An overview of 2nd order random walk in node2vec. The walk has transitioned to node g from node e. Next, it is assessing the next node in the walk after node g. Labels on the edges denote the transition probability from g to its neighbors

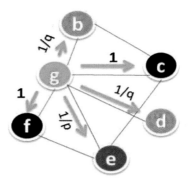

- p controls the probability of immediately sampling an already visited node. A smaller value of p makes sure that it is highly likely to revisit a node in the next two steps and walk remains close to the start (root) node. On the contrary, it avoids redundancy in 2-hops if p is fixed to a high value.
- On the other hand, q allows the walk to differentiate between the inward and the outward nodes. If $q > 1$, then the walk is likely to visit nodes near the node z while if $q < 1$, then walk is biased to the nodes farther from node z. Hence, node2vec can approximate BFS behavior when $q > 1$, which motivates the inward exploration, while DFS like exploration when $q < 1$, which motivates the outward movement.

So, node2vec samples multiple neighborhood sets $\mathcal{N}_i$ of fixed size for each source node v_i using the second order-random walk strategy. node2vec then follows the skip-gram model, and similar to DeepWalk, node2vec maximizes the log-likelihood of visiting the neighborhood $\mathcal{N}_{v_i}$ of v_i given the representation of node v_i described by mapping g.

Further, to make the optimization manageable, node2vec makes some assumptions.

- First, the conditional independence assumption, which factorizes the log-probability by assuming that the likelihood of observing a context node does not depend on the knowledge of other context nodes, conditioned on the representation of node v_i.
- Second, the symmetry in the feature space states that the source and the context nodes have a symmetric impact on each other in the feature dimension.

Hence, node2vec models the log-probability of each node-context pair as a soft-max unit. This unit is described by the dot product between their features. The following equations summarize the optimization problem:

$$\min_{g} - \sum_{v_j \in \mathcal{N}_{v_i}} \log P(v_j | g(v_i)) \quad where,$$

$$P(v_j | g(v_i)) = \frac{\exp(g(v_i).g(v_j))}{\sum_{v_k \in V} \exp(g(v_k).g(v_i))} \tag{4.3}$$

In the above equations, $\mathcal{N}_{v_i}$ is the neighborhood set of node v_i and V is the set of all the nodes. Moreover, $\sum_{v_k \in V} \exp(g(v_k).g(v_i))$ is infeasible to calculate, and hence $P(v_j | g(v_i))$ is approximated by using the negative sampling technique as is done in word2vec.

4.3 Matrix Factorization Based Algorithms

4.3.1 Network Representation Learning with Rich Text Information

Most network representation learning (NRL) algorithms consider the network structure, but they do not integrate node attributes. A trivial way is to learn representations from node features and the network structure independently and then merge them to form a single representation. However, this approach will lose the impact of the important interactions between node attributes and the network structure. Text-associated DeepWalk ($TADW$) architecture provides a solution for this and integrates the structure and the text features of the nodes in the representation learning under matrix factorization.

Let $G = (V, E)$ be a graph with a set of N vertices V and a set of edges E. Suppose, $A \in \mathcal{R}^{N \times N}$ is the normalized adjacency matrix of graph G such that if $(i, j) \in E$ then A_{ij} is equal to the 1/(degree of node i) and if $(i, j) \notin E$ then $A_{ij} = 0$.

- $TADW$ first proves the equivalence between the DeepWalk and matrix factorization. Let $P \in \mathcal{R}^{N \times N}$ be a matrix where P_{ij} is the log of the average probability, i.e., the probability of node i visiting node j randomly in a fixed number of steps(t).

  - DeepWalk factorizes P into two matrices $Z \in \mathcal{R}^{k \times N}$ and $S \in \mathcal{R}^{k \times N}$, where each column of Z is the k-dimensional representation of node and each column of S is the k-dimensional representation of the context node and further $k \ll N$. DeepWalk aims to solve Z and S to minimize:

$$\min_{Z,S} ||P - Z^T S||_F^2 + \frac{\lambda}{2}(||Z||_F^2 + ||S||_F^2) \tag{4.4}$$

- Next, TADW is proposed, aiming to integrate node features with the network structure to learn enhanced node representations using inductive matrix completion.
- So, TADW aims to solve Z and S to minimize the following:

$$\min_{Z,S} ||P - Z^T S B||_F^2 + \frac{\lambda}{2}(||Z||_F^2 + ||S||_F^2) \tag{4.5}$$

Here, $B \in \mathcal{R}^{D \times N}$ is the matrix of the original D dimensional node features.

- In DeepWalk, the ijth entry of matrix P is $P_{ij} = \log([e_i(A + A^2 + \cdots + A^t)]_j/t$. But in TADW, the matrix $P = \frac{(A+A^2)}{2}$ is factorized as there are more non-zero entries in $\log P$, and the complexity of the matrix factorization with square loss increases in proportion to the number of non-zero elements in the matrix.
- Both matrices $Z \in \mathcal{R}^{k \times N}$ and SB are considered as the low dimensional node representations, and in TADW, they are concatenated together to form the final $2K$ dimensional node representations.

4.3.2 GraRep: Learning Graph Representations with Global Structural Information

Network representation learning schemes discussed so far do not capture k-step ($k > 2$) relations between nodes. GraRep is based on the view that the k-step relations between nodes are essential to capture the global structural information of the graph.

- GraRep encodes k-step relations between nodes, with various values of k, by exploiting multiple higher-order transition matrices associated with the graph and defines distinct loss functions for encoding various k-step relations.
- GraRep aims at generating a global representation matrix $W \in \mathcal{R}^{N \times F}$ where F is the feature dimension, and ith row of W denotes the representation of node i.
- These embeddings encapsulate the respective graph's global structural details by examining different connections of the nodes in respect of different transition steps and encoding long-range relationships between nodes.

- **Transition Matrix:**

  - Given nodes i and j, GraRep aims to get global representations that encode the distant relations between these nodes, which requires studying the strength of their relations.
  - For this, it is checked to see if a path exists between the nodes and then compute the probability of transition $(p_k(j|i))$ from node i to node j in k steps.
  - The 2-step transition probability matrix is $A^2 = A.A$, 3-step transition matrix is $A^3 = A.A.A$, and likewise, k-step transition matrix is A^k. The probability of k-step transition from node i to node j $(p_k(j|i))$ is given by the ijth entry of A^k i.e., A_{ij}^k.

- **Loss Function:**

  - For each k, GraRep first samples all paths of k steps which start with i and end at j.

- The goal is to maximize the probability of all such ij pairs that belong to graph G and minimize the probability of all the other pairs which are not from the graph.
- Inspired by the skip-gram model, the NCE (Noise Contrastive Estimation) is exploited to define the objective function.
- For each $k \in \{1, \ldots, K\}$, GraRep defines the k-step loss function for the complete graph as:

$$L_k = \sum_{i \in V} L_k(i) \quad where,$$

$$L_k(i) = \left(\sum_{j \in V} p_k(j|i) \log \sigma(\overrightarrow{i} . \overrightarrow{j}) \right) + \gamma \mathbb{E}_{j' \sim p_k(V)} [\log \sigma(-\overrightarrow{i} . \overrightarrow{j'})] \quad (4.6)$$

Here $p_k(i|j)$ is the probability of transition from node i to node j in k-steps, σ is a sigmoid activation function, hyperparameter γ is the number of negative samples. $\mathbb{E}_{j' \sim p_k(V)}$ is the expectation when j' follows the distribution $p_k(V)$, where j' is a negative sample.

- For large k, transition probabilities converge to a particular value and hence, $p_k(j)$ is calculated as:

$$p_k(j) = \sum_{i'} q(i') p_k(j|i') \quad (4.7)$$

Here $q(i')$ denotes the likelihood of node i' being the starting node which is $\frac{1}{N}$ as it is assumed to follow a uniform distribution.

Matrix Factorization Based Optimization: GraRep follows the matrix factorization version of the skip-gram model. But GraRep encodes the higher-order proximity between two nodes such that nodes having common neighbors at k-step ($1 \leq k$) will have similar embeddings.

For each $k \in \{1, \ldots, K\}$, GraRep describes the context nodes as the k-step neighbors and performs a three step process to learn k-step representations for all the nodes.

- First, GraRep calculates a k-step transition probability matrix T^k.
- Second, it computes the k-step representation, $\forall k = 1, \ldots, K$. It uses SVD to factorize the log probability matrix as follows:

$$X^k = S^k \sigma^k (W^k)^T$$
$$X^k \approx S^k_d \sigma^k_d (W^k_d)^T \quad (4.8)$$

Here S^k_d and W^k_d are the d columns of the respective matrices; σ^k_d are the top d singular values and the matrix $S^k_d \sqrt{(\sigma^k_d)}$ denotes the node representations.

- Lastly, it concatenates all the learned k-step representations to get a final representation for each node.

4.4   Graph Neural Networks

Representation learning is vigorously impacted by the advent of deep learning, which led to the introduction of Graph neural networks ($GNNs$). $GNNs$ have gained significant attention for node representation and classification tasks.

Graphs belong to a non-Euclidean space and are irregular structures with an unordered sequence of nodes. Further, nodes in a graph are dependent on each other because of the existing relations (edges). These properties foist problems for downstream tasks, and unlike images or text, machine learning tools cannot be used directly on graphs. GNNs map nodes from non-Euclidean space to Euclidean space by leveraging both the node attributes and the structure.

Given a graph $G = (V, E)$, where V denotes the set of nodes and E denotes the edge set, $GNNs$ learn node embeddings by aggregating node features from the neighbors of the node as:

- First, $AGGREGATE$ function computes the aggregation of the embeddings of neighbors of node i and outputs a single vector u_i^{k+1} as described below:

$$u_i^{k+1} = AGGREGATE^k \left(\left\{ h_j^k : j \in \mathcal{N}_{(i)} \right\} \right) \tag{4.9}$$

Here h_i^k describes the output embeddings of node i in the kth GNN layer, and $\mathcal{N}_G(i)$ is the neighbor set of node i. The $AGGREGATE$ function should be permutation invariant because neighbors of nodes are not in a consistent order across all the nodes and should also work with variable neighborhood sizes.
- Finally, the $COMBINE$ function combines the embedding vector of the ith node from kth layer and the aggregated representation vector of the neighbors of node i (u_i^{k+1}) to output an updated embedding vector of node i in $k + 1$th GNN layer as described by the equation below:

$$h_i^{k+1} = COMBINE^l \left(h_i^k, u_i^{k+1} \right) \tag{4.10}$$

In this section, we discuss some state-of-the-art GNN based algorithms for learning node representations.

4.4.1   Semi-Supervised Classification with Graph Convolutional Networks

Graph Convolutional Network generalizes the convolution operator to the graph domain. Formally, graph convolutional network (GCN) computes a weighted mean of the representations of the neighbors of a node to find the representation of the node. It performs in a transductive setting and needs the complete train and test sets during the learning stage.

- GCN takes the following as input for a graph $G = (V, E)$, where V is the set of nodes and E is the set of edges:

  - An adjacency matrix $A \in \mathbb{R}^{N \times N}$ (where N is the number of nodes $|V|$) describing the structure of the graph.
  - A node feature matrix $X \in \mathbb{R}^{N \times D}$ where N is the number of nodes and D is the feature dimension. X_i denotes the feature vector of node i.

- Given these inputs, the layer-wise propagation rule (lth layer) of graph convolution can be defined as:

$$H^{l+1} = \sigma(\hat{D}^{-\frac{1}{2}} \hat{A} \hat{D}^{-\frac{1}{2}} H^l W^l) \tag{4.11}$$

Here $H^l \in \mathcal{R}^{N \times F'}$ is the node representation matrix in the lth GCN layer with D feature dimension. $\hat{A} = A + I_N$ denotes the adjacency matrix with a self loop at each node of G. $\hat{D} \in \mathbb{R}^{N \times N}$ is the diagonal degree matrix, and ith diagonal entry is $\hat{d}_i = \sum_{j \in \{1,2,...,N\}} \hat{A}_{ij}$. $W^l \in \mathbb{R}^{F' \times F'}$ is a convolution matrix with feature dimension F' (except $W^0 \in \mathbb{R}^{D \times F'}$), which is shared across all the nodes of the graph. $\sigma()$ is an activation function such as ReLU and the initial node representation matrix $H^0 = X$. The final node representation matrix Z is equal to the output representation matrix H^L in the last GCN layer L.

- At last, one more GCN layer with convolution matrix $W^P \in \mathcal{R}^{F' \times C}$, where C is the number of distinct node labels, and with an activation function σ that is set to softmax is added to generate predictions $Y \in \mathcal{R}^{N \times C}$ for node classification task.
- Finally, the cross-entropy loss is calculated to perform semi-supervised multi-class classification as:

$$\mathbb{L} = \sum_{i \in N^L} \sum_{j=1}^{C} Y_{ij} log O_{ij} \tag{4.12}$$

Here Y and O are the predicted and actual labels sets and N^L contains the indices of labeled nodes.

Further, this propagation rule solves two issues.

- First, it adds self-loops in matrix A so that the computation of a node embedding considers the node features along with its neighbor's features. Thus, Eq. 4.11 computes the new representation matrix in layer $l + 1$, denoted by H^{l+1}, as the aggregation of the embeddings of the neighbors and the node itself.
- Second, matrix A is normalized by the degree matrix D. This makes the scale of the features of all the nodes the same because the sum of the degrees for each node is equal to 1 after normalization.

4.4.2   Graph Attention Network

Another GNN based supervised model to learn the representation of nodes is Graph Attention Network (GAT), which uses attention in the GCN framework.

- GAT leverages self-attention on the graph domain to learn the importance of a node in determining the label of another node. More precisely, GAT captures the significance of a node to the embedding of another node in its neighborhood by training an attention vector.
- The following equations compute the attention scores for node pairs:

$$c_{pq} = \text{LeakyReLU}\left(a^T[\mathbf{W}z_p \| \mathbf{W}z_q]\right)$$
$$\beta_{pq} = \text{softmax}_q\left(\exp(c_{pq})\right)$$
$$= \frac{\exp\left(\text{LeakyReLU}\left(a^T[\mathbf{W}z_p \| \mathbf{W}z_q]\right)\right)}{\sum\limits_{i \in N_p} \exp\left(\text{LeakyReLU}\left(a^T[\mathbf{W}z_p \| \mathbf{W}z_i]\right)\right)} \tag{4.13}$$

  - In these equations, $z_q \in \mathcal{R}^k$ is the feature vector of qth node with feature dimension k, $\mathbf{W} \in \mathcal{R}^{k \times k'}$ is a trainable weight matrix where k' is the output feature dimension, $\|$ denotes the concatenation, and N_p denotes the set of immediate neighbors of node p, including node p itself; $a \in \mathcal{R}^{2k'}$, which is a weight vector of a single layer neural network, and LeakyReLU non-linearity, are used to learn the attention mechanism which computes the attention coefficient c_{pq}.
  - The attention coefficient c_{pq} denotes the importance of the features of node q to node p. c_{pq} is normalized across the neighbors ($\forall q \in N_p$) of pth node to make the attention coefficients comparable across all the nodes. These normalized scores are denoted by β_{pq}.
  - Also, the attention coefficients c_{pq} are computed only for nodes $q \in N_p$ (only for those pairs of nodes that are first-order neighbors), i.e., GAT performs masked attention, henceforth, inducing the structure of the graph.

- These learned normalized attention values are further used to calculate the linear combination of the corresponding feature vectors to calculate the new output features of each node as follows:

$$z'_p = \sigma\left(\sum_{q \in N_p} \beta_{pq} \mathbf{W} z_q\right) \tag{4.14}$$

Here β_{pq} is the attention score calculated in Eq. 4.13, σ is the non-linearity, and $z'_p \in \mathcal{R}^k$ is the new output feature vector of node p. Until here, we have discussed a single head GAT layer, which can also be extended to multi-head attention by learning r different attention mechanisms with separate linear transformation weights and parameter a for each head.

– For multi-head attention GAT layer, the output features of node p are computed as:

$$z'_p = \Big\|_{r=1}^{r=R} \sigma\Big(\sum_{q \in N_p} \beta^r_{pq} \mathbf{W}^r z_q\Big) \tag{4.15}$$

Here, W^r is the transformation weight matrix for head r, β^r_{pq} is the normalized attention score determined by the attention head r, and $\|$ is the concatenation of the features from all the heads to learn the final feature vector z'_p.

– Moreover, if the last GAT layer (i.e., prediction layer) also computes multi-head attention, then instead of concatenation, the sum operator is used to calculate z'_p as follows:

$$z'_p = \sigma\Big(\frac{1}{r}\sum_{r=1}^{R}\sum_{q \in N_p} \beta^r_{pq} \mathbf{W}^r z_q\Big) \tag{4.16}$$

Here σ is either the sigmoid or softmax non linearity to convert the output of this layer into predictions, and $W^r \in \mathcal{R}^{N \times C}$ with C distinct node labels is the transformation weight matrix for head r in the prediction layer.

4.4.3   Inductive Representation Learning on Large Graphs (GraphSAGE)

GraphSAGE was introduced as an improvement over GCN. Many GNN based approaches are transductive, i.e., all node features and their corresponding connections are leveraged during training, which implies that for new nodes, the model will have to be re-trained. But GraphSAGE introduces an inductive framework that, instead of learning representations for each node separately, learns a set of aggregator functions that sample and aggregate the information from the local neighborhood of a node to learn the node's representations. During the testing stage, these trained aggregator functions are used to generate representations for the new nodes.

Let $G = (V, E)$ be a graph with a set of N nodes V and a set of edges E.

• GraphSAGE combines information present in the node and the node's local structure for generating new latent representation as follows:

$$z^t_{N_i} = AGGREGATE_t\big(\{z^{t-1}_j, \forall j \in N_i\}\big)$$
$$z^t_i = \sigma\big(\mathbf{W}^t.[z^{t-1}_i \| z^t_{N_i}]\big) \tag{4.17}$$

– Here N_i denotes a fixed size set of uniformly sampled immediate neighbors of node i.
– $AGGREGATE_t$ function, $\forall t \in \{1, \ldots, T\}$, aggregates embeddings of the immediate neighbors of the node, $\{z^{t-1}_j, \forall j \in N_i\}$, and generates vector $z^t_{N_i}$ representing the neighborhood of node i.

- The aggregation step operates on embeddings from previous layer, such as $AGGREGATE_t$ uses embeddings from layer $t-1$.
- At layer $t = 0$, these embeddings are the input node features.

- Next, GraphSAGE concatenates the neighborhood vector, $z_{N_i}^t$, with the embedding vector from the previous layer of node i, z_i^{t-1}.
- Finally, the embedding of node i in layer t is computed by linearly transforming this concatenated vector using weight matrix W^t and applying a nonlinear activation function σ on this transformed vector.

  - These weight parameters W^t transmit features between different layers (or different depths or different $AGGREGATE$ functions). The generated node embeddings are used by the next layer, i.e., $AGGREGATE^{t+1}$ function.

- To learn neighborhood aggregation, GraphSAGE introduces three aggregator functions: mean aggregator, pooling based aggregator, and LSTM aggregator.

  - The mean aggregator takes the mean of the neighbor's embeddings to derive a representation of the node's local neighborhood structure.
  - LSTM aggregator applies LSTM to a random permutation of the representations of the node's neighbors to generate a representation of the node's neighborhood.
  - The pooling aggregator passes feature vectors of a node's neighbors through a neural network and aggregates these transformed features using the max-pooling operator.

- Further, GraphSAGE introduces a graph-based unsupervised loss function (discussed in the Eq. 4.18 below), making nearby nodes identical and distant nodes highly dissimilar in the embedding space and operates on the last layer output embeddings, i.e., z_i^T, $\forall i \in V$. The parameters of the model are tuned using stochastic gradient descent.

$$J_G(Z_i) = -log\left(\sigma(z_i^T z_j)\right) - S_n . E_{j_n \sim P_s(j)} \log\left(\sigma(-z_i^T z_{j_n})\right) \qquad (4.18)$$

Here, j is a node that co-occurs near node i in a random walk of fixed length. z_i and z_j are the representations of nodes i and j in the last layer T, S_n is the number of negative samples, P_s is the distribution for negative sampling, and σ is the sigmoid activation function.

4.4.4 Jumping Knowledge Networks for Node Representations

This work manages how to specifically leverage information from neighborhoods of different ranges.

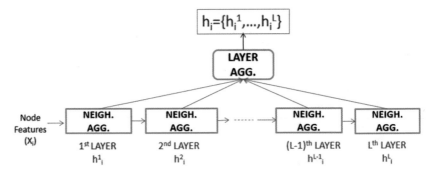

Fig. 4.3 The architecture of JK-Net. NEIGH. AGG. denote the neighborhood aggregation. X_i is the input feature vector, and h_i is the final feature vector of node i

- A node representation depends on different ranges of neighborhoods, depending on the structure of the graph.
- The introduced JK-nets are adaptive to the local neighborhood properties, i.e., for each node, different ranges of neighborhoods (or influence distribution of a node) are exploited to learn structure-aware embeddings.
- In any aggregation based method, as discussed in GCN and GAT, each layer increases the radii of influence distribution by aggregating information from the prior layer.
- Large radii aggregate too much information, whereas small radii aggregate inadequate information. As depicted in Fig. 4.3, at each layer l, new features h_i^l of node i are computed, and as l increases, h_i^l has more information for larger neighborhoods centered around i.
- Thus, JK-Net combines information from different locality neighborhoods leveraging skip connections from all the layers to the last layer and selective but adaptive attention mechanism. Further, these models include various aggregations at the last layer of JK-Net to selectively combine information for each node.
- As Fig. 4.3 describes, in JK-Net, all layers' outputs are skipped to the last layer, where Jk-Net applies layer aggregation (concatenation, LSTM) to select different neighborhood ranges for each node.
- Thus, at the last layer, JK-Net chooses from all the intermediate embeddings that jump to the layer for every node. This is done independently for each node, and thus, JK-Net adaptively selects the effective size of the neighborhood on which the node's representation depends.
- Let $\{h_i^1, h_i^2, \ldots, h_i^K\}$ be the representations of node i (node i's jumping representations) at all layers, $1, 2, \ldots, K$. JK-Nets have three different ways to aggregate the jumping representations, i.e., to perform layer aggregation: (1) Concatenation, (2) Max-Pool, and (3) LSTM-attention.

  1. In concatenation, the final vector of a node i is formed by combining representations from all the layers $[h_i^1, h_i^2, \ldots, h_i^K]$ and applying a linear transformation.

But if the transformation weights are shared across all the nodes, then this operator is not node-adaptive.

2. Max-Pooling selects element-wise maximum $max(h_i^1, h_i^2, \ldots, h_i^K)$. For every feature dimension, it picks the most significant layer. Thus, this operation is adaptive and also does not include any additional trainable parameters.

3. In LSTM-attention, $\{h_i^1, h_i^2, \ldots, h_i^K\}$ are fed to a bi-directional LSTM, which returns two latent representations (forward and backward LSTM), denoted by x_i^k and y_i^k, respectively, $\forall k \in \{1, \ldots, K\}$. Next, a scalar r_i^k is generated for each kth layer by linearly mapping the concatenated features $[x_i^k || y_i^k]$. These scores are normalized by applying softmax on $\{r_i^1, r_i^2, \ldots, r_i^K\}$ to determine the importance of node i on the neighborhoods of various ranges. Finally, the embedding of node i is formed by taking a weighted sum of $[x_i^k || y_i^k]$ with weights being the normalized scores. This operation is too adaptive as attention scores vary for each node.

- JK-Nets can further be employed in other models such as GraphSAGE, Graph Convolutional Networks, and Graph Attention Networks for performance enhancement.

4.4.5 Deep Graph Infomax

Deep graph Infomax (DGI) is an unsupervised learning algorithm, based on an objective of mutual information maximization for node representation learning.

Suppose $G = (V, E)$ is the input graph with a set of N nodes V and a set of edges E. G is augmented with the node features matrix $X \in \mathcal{R}^{N \times F}$, where F is the feature dimension and an adjacency matrix $A \in \mathcal{R}^{N \times N}$.

- DGI aims to learn an encoder, $\mathbb{E} : \mathcal{R}^{N \times F} \times \mathcal{R}^{N \times N} \hookrightarrow \mathcal{R}^{N \times F'}$, such that $\mathbb{E} = Z$, where $Z \in \mathcal{R}^{N \times F'}$ denotes the node (i.e., local) representations matrix with $z_i \in \mathcal{R}^{F'}$ representing node i.

  - The encoder is GCN encoder for transductive setting as shown in the below equation:

  $$E(X, A) = \sigma(\hat{D}^{-\frac{1}{2}} \hat{A} \hat{D}^{-\frac{1}{2}} X W) \tag{4.19}$$

  - The encoder is a mean-pooling propagation rule (GraphSAGE-GCN) for inductive learning and is described as follows:

  $$E(X, A) = \sigma(\hat{D}^{-1} \hat{A} X W) \tag{4.20}$$

  In both these Eqs. (4.19 and 4.20), $\hat{A} = A + I_N$, i.e., the adjacency matrix after adding a self loop to each node of G. $\hat{D} \in \mathbb{R}^{N \times N}$ is the diagonal degree matrix, σ is ReLU activation, and $W^l \in \mathbb{R}^{F \times F'}$ is a trainable parameter matrix, where F is the input dimension, and F' is the output dimension.

As the node representations, $\{z_i, \forall i \in \{1, \ldots, N\}\}$, are generated by aggregating information in the node's local neighborhood and can outline a patch centered at the ith node, thus are also referred to as patch representations.

- To learn the encoder, DGI maximizes the local/global mutual information. Formally, the aim is to learn the node embeddings that encode global information about the whole graph. DGI uses a READOUT layer $\mathbb{Q} : \mathcal{R}^{N \times F} \longrightarrow \mathcal{R}^F$ to learn a graph representation, denoted by a summary vector s, from node representations that captures global information for the whole graph.

  - In DGI, the READOUT layer is a simple averaging of all the patch (node) representations.

$$\mathbb{Q}(Z) = \sigma \sum_{i=1}^{N} z_i \qquad (4.21)$$

- These operations can be summarized as $s = \mathbb{Q}(\mathbb{E}(X, A))$.
- As a substitute for MI maximization, DGI introduces a discriminator $T : \mathcal{R}^F \times \mathcal{R}^F \longrightarrow \mathcal{R}$, where the ith entry, $T(z_i, s)$, denotes the score given to the ith patch representation (z_i) and summary vector (s) pair. A higher probability score implies the corresponding patch is contained within the summary.

  - For discriminator T, the negative samples (represented by z'_i) are made by pairing s of graph (X,A) with node representations of a different graph (X', A') from the graph collection dataset. When the input is a single graph, the negative samples are generated by a corruption function $\mathbb{K} : \mathcal{R}^{N \times F} \times \mathcal{R}^{N \times N} \longrightarrow \mathcal{R}^{M \times F} \times \mathcal{R}^{M \times M}$ such that $(X', A' = \mathbb{K}(X, A))$.
  - For discriminators, DGI uses a simple and effective bilinear scoring function. That is,

$$T(z_i, s) = \sigma(z_i^T W s) \qquad (4.22)$$

Here σ is a sigmoid activation function that converts scores into the probabilities, and W is a parameter matrix.

- DGI uses a noise contrastive type objective with a standard binary cross-entropy loss function between the positive samples and the negative samples. The objective function is explained as:

$$\mathcal{L} = \frac{1}{N + M} \left(\sum_{i=1}^{N} \mathbb{E}_{(X,A)} \left[log\ T\left(z_i, s \right) \right] \right.$$

$$\left. + \sum_{j=1}^{M} \mathbb{E}_{(X',A')} \left[log\ \left(1 - T\left(z'_i, s \right) \right) \right] \right) \qquad (4.23)$$

- Consequently, all the patch representations preserve mutual information with the summary vector, which further maintains the similarities between patches.

To summarize:

- First, the negative samples are generated using the corruption function such that $(X', A') = \mathbb{K}(X, A)$.
- Second, patch representations are learned by using the encoder on the input graph, $Z = \mathbb{E}(X, A)$. Similarly, patch representations for the negative samples are computed, $Z' = \mathbb{E}(X', A')$.
- Following this, a summary vector s for the input graph is generated using the READOUT layer on the learned patch representations Z.
- Finally, parameters of the model are trained by applying gradient descent.

4.5 Experimental Evaluation

As stated earlier, these learned node representations are used for various network analysis tasks such as node classification, node clustering, link predictions, etc. Any traditional ML tool can be used for these downstream tasks using the learned embeddings as the input features. Further, the performance on these tasks can be used to evaluate the quality of these learned embeddings or the learning algorithms' efficacy.

In this section, we compare the results of the discussed algorithms on citation networks i.e., cora, citeseer, and pubmed on node classification and node clustering tasks. To give more insights into the performance of these algorithms, we further show t-SNE visualization of the representations generated by these algorithms.

4.5.1 Node Classification

A standard split is introduced for node classification in GCN. We report the results for all the algorithms on the same set-up to have a fair comparison between different representation learning algorithms. Nodes of a graph are divided into train, validation, and test sets as follows:

- 20 nodes are randomly selected from each class as labeled nodes for the training set, 100 nodes are used for the test set, and 500 nodes are selected for the validation set for hyper-parameter optimization.
- The Cora dataset has seven distinct node labels; therefore, the sizes of train/val/test sets are 140/500/100. The Citeseer dataset has six distinct classes; thus, the sizes of train/val/test sets are 120/500/100, and the PubMed dataset has three distinct classes; hence, the sizes of train/val/test sets are 60/500/100.

Table 4.1 Classification accuracy (%) of different algorithms for node classification

Dataset	Cora	Citeseer	Pubmed
DeepWalk	67.3	41.50	66.4
node2vec	70.8	47.90	71.80
TADW	71.40	58.20	70.00
GraRep	70.50	46.90	70.80
GCN	81.5	70.3	**79.0**
GAT	**83.0±0.7**	72.5±0.7	**79.0±0.3**
GraphSAGE	82.9	71.4	78.60
JK-Net	82.7±0.4	**73.0±0.5**	77.9± 0.4
DGI	82.3±0.6	71.8±0.7	76.8±0.6

4.5.1.1   Performance Analysis

Table 4.1 shows the classification accuracy of seven algorithms for node classification on cora, citeseer and pubmed datasets.

- We can observe that on Cora and PubMed datasets, GAT is outperforming all unsupervised and supervised algorithms. Whereas on the citeseer dataset, JK-Net achieves the best performance. Also, JK-Net is highly competitive to GAT on cora dataset with performance gap of less than 1%.
- Further, it is interesting to note that other supervised algorithms (GCN and Graph-SAGE) are also able to outperform the random walk and the matrix factorization based unsupervised approaches on all the datasets but are outperformed by DGI (an information maximization based unsupervised approach) on some datasets.
- This observed performance gain in supervised algorithms over unsupervised algorithms depicts the importance of leveraging label information during the training stage.
- Moreover, Deep Graph Infomax (DGI, an unsupervised algorithm) is performing better than all other unsupervised algorithms.

4.5.2   Node Clustering

We further demonstrate the performance of the unsupervised algorithms on the node clustering task.

- We use the same embeddings which are used for the node classification task.
- The learned node embeddings from any algorithm are fed to the K-Means algorithm, which returns the clusters.
- We use the unsupervised clustering accuracy to assess the quality of the generated clusters. The unsupervised clustering metric can be defined as below:

Table 4.2 Clustering accuracy (%)

Algorithms	Cora	Citeseer	Pubmed
DeepWalk	59.84	45.68	67.15
node2vec	62.14	42.76	**68.21**
TADW	67.78	**68.42**	60.79
GraRep	49.23	36.49	58.02
DGI	**69.01**	67.48	51.52

$$Clustering_A ccuracy(\hat{\mathcal{L}}, \mathcal{L}) = \max_Q \frac{\sum_{i=1}^{N} \mathbf{1}(Q(\hat{\mathcal{L}}_i) = \mathcal{L}_i)}{N} \qquad (4.24)$$

Here, $\mathcal{L}_i$ represents the ground truth label of sample point i and N is the number of samples. The permutation of labels is denoted by Q, and $\hat{\mathcal{L}}$ symbolizes the output assignments of an algorithm, i.e., the clustering assignments. $\mathbf{1}$ denotes the logical operator that returns 1 if the argument is true otherwise 0.

• The unsupervised clustering accuracy metric uses different orderings of the labels and selects the label permutation which outputs the best accuracy for the clustering task.

4.5.2.1 Performance Analysis on Clustering Task

Table 4.2 shows the clustering accuracy of the unsupervised algorithms for clustering task on three real-world datasets (cora, citeseer and pubmed).

• We can observe that DGI is outperforming all algorithms on cora dataset.
• While on citeseer dataset, TADW is giving state-of-the-art performance, and on pubmed dataset, node2vec performs the best among the selected algorithms.
• Further, DeepWalk is competitive to node2vec on pubmed dataset, but node2vec outperforms DGI with a good margin. In contrast, DGI is able to perform well on the citeseer dataset with <1% performance gap with the best algorithm.

4.5.3 Visualization

We further show the visualization of the nodes as another downstream task to assess the quality of the node representations.

• We fed the learned node embeddings to t-SNE, which outputs the 2-D embeddings.

- For this task, we use the same embeddings generated in the node classification task for unsupervised algorithms, and we keep the hyperparameters' tuning fix as in node classification for supervised algorithms to generate the node representations.
- We select a subset of approaches for the visualization task. Figures 4.4 and 4.5 show the t-SNE visualizations of the nodes of Cora and Citeseer datasets by the embeddings learned by different algorithms, respectively.
- We select three unsupervised algorithms; in Figs. 4.4a and 4.5a, we use the node2vec model. In Figs. 4.4b and 4.5b, we pick node representations generated by TADW, and in Figs. 4.4c and 4.5c, we pick DGI algorithm.

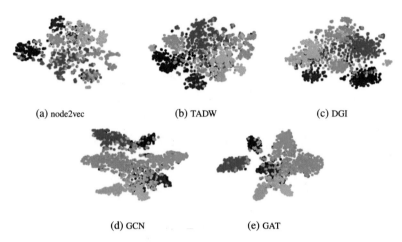

(a) node2vec   (b) TADW   (c) DGI

(d) GCN   (e) GAT

Fig. 4.4 t-SNE visualization of cora dataset (different colors depicts different node labels) by the embeddings produced by different algorithms

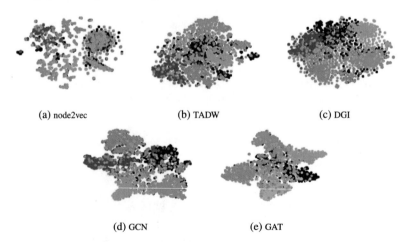

(a) node2vec   (b) TADW   (c) DGI

(d) GCN   (e) GAT

Fig. 4.5 t-SNE visualization of citeseer dataset (different colors depict different node labels) by the embeddings produced by different algorithms

- Further, we also select two supervised algorithms; Figs. 4.4d and 4.5d depict the t-SNE visualizations by the embeddings generated by GCN, and Figs. 4.4e and 4.5e depict the performance of embeddings learned by the GAT algorithm.
- In these plots, different colors denote separate node labels. The performance is optimal when the different colors are separated the most, i.e., generate distinct clusters in the plot.

4.5.4 Performance Analysis

- We can observe that the best performance on Cora dataset is achieved in Fig. 4.4e in which the representations are generated using GAT.
- Also, the performances of the supervised algorithms are better than the unsupervised algorithms.
- On the other hand, comparing only unsupervised algorithms, performance in Fig. 4.4c is better than in Figs. 4.4a and 4.5a for Cora dataset.
- A similar observation of the enhanced performance of GAT over all algorithms (unsupervised and supervised) and the improved performance of DGI over other unsupervised algorithms are also prominent in Fig. 4.5 on citeseer dataset.

Bibliography

1. Hamilton WL, Ying R, Leskovec J (2017) Representation learning on graphs: methods and applications. arXiv preprint arXiv:1709.05584
2. Zhang D, Yin J, Zhu X, Zhang C (2018) Network representation learning: a survey. IEEE Trans Big Data
3. Kipf TN, Welling M (2016) Semi-supervised classification with graph convolutional networks. arXiv preprint arXiv:1609.02907
4. Veličković P, Cucurull G, Casanova A, Romero A, Lio P, Bengio Y (2017) Graph attention networks. arXiv preprint arXiv:1710.10903
5. Hamilton W, Ying Z, Leskovec J (2017) Inductive representation learning on large graphs. In: Advances in neural information processing systems, pp 1024–1034
6. Xu K, Li C, Tian Y, Sonobe T, Kawarabayashi KI, Jegelka S (2018) Representation learning on graphs with jumping knowledge networks. arXiv preprint arXiv:1806.03536
7. Veličković P, Fedus W, Hamilton WL, Liò P, Bengio Y, Hjelm RD (2018) Deep graph infomax. arXiv preprint arXiv:1809.10341
8. Grover A, Leskovec J (2016) node2vec: scalable feature learning for networks. In: Proceedings of the 22nd ACM SIGKDD international conference on knowledge discovery and data mining, pp 855–864
9. Perozzi B, Al-Rfou R, Skiena S (2014) Deepwalk: online learning of social representations. In: Proceedings of the 20th ACM SIGKDD international conference on knowledge discovery and data mining, pp 701–710

10. Yang C, Liu Z, Zhao D, Sun M, Chang EY (2015) Network representation learning with rich text information. In: IJCAI, vol 2015, pp 2111–2117
11. Cao S, Lu W, Xu Q (2015) Grarep: learning graph representations with global structural information. In: Proceedings of the 24th ACM international on conference on information and knowledge management, pp 891–900

Chapter 5
Embedding Graphs

Graphs are of vital importance in many application domains. Different components of a graph are embedded in the vector space. The previous chapter discussed some node embedding algorithms and presented comparisons based on empirical results.

In this chapter, we consider graph level representations. Graph embedding aims at embedding the entire graph in a low-dimensional vector space. Embeddings are such that the properties of the whole graph are captured using similarity between graphs. These representations are used for many graph level analysis tasks, including graph classification, graph visualization, etc. They have many critical applications, such as finding anti-cancer activity, finding the molecule toxicity level, and many more.

Many graph kernels were introduced to map graphs in to the Hilbert space to preserve the graphs' implicit/explicit properties.

- They compute the similarity between graphs in the Hilbert space. They have attained state-of-the-art performance on various datasets.
- To learn the similarity between graphs, they compare the substructures of the respective graphs. Different graph kernels were proposed depending on the substructures such as the Kernels based on rooted subtrees, the random walk based kernel, the shortest path based kernels, the graphlet counting based kernel, and the WL isomorphism test based kernel known as Weisfeiler-Lehman subtree kernel.
- However, all these kernels exploit the handcrafted features, and hence they are not skilled in generalizing to the data.

Recently, many pooling techniques and GNNs are introduced to learn graph vectors. For instance, Infograph is a recently introduced an unsupervised algorithm that uses mutual information maximization to learn graph embeddings. This chapter discusses some of these state-of-the-art algorithms for embedding graphs into a vector space, i.e., graph representation learning algorithms.

Graph Pooling: Earlier, trivial pooling operations are being used, for instance, first learning node representations and applying aggregation operations (max, sum,

© The Author(s), under exclusive license to Springer Nature Singapore Pte Ltd. 2021
M. Aggarwal and M. N. Murty, *Machine Learning in Social Networks*,
SpringerBriefs in Computational Intelligence,
https://doi.org/10.1007/978-981-33-4022-0_5

and mean) on these learned vectors to generate a representation of the entire graph. But these operations destroy the graph structure and cannot be used for graphs with varying structures. Recently, many trainable pooling mechanisms are proposed which learn structure or feature dependent clusters to form a coarser version of the graphs. There are two types of graph pooling techniques, global pooling and hierarchical pooling:

- **Hierarchical pooling:** This technique learns a hierarchical representation of the graph by recursively mapping the graph into a smaller graph and maintaining the implicit hierarchical structure of the graph.
- **Global pooling:** This technique directly pools the representations of all the nodes of a graph by using some simple aggregation to generate a single representation of the graph, irrespective of the graph structure, and hence is inherently flat. This works better when aggregating all the nodes using a single function is a more suitable option.

5.1 SortPool

Graphs, unlike images, do not have an ordered sequence of nodes; consequently, the neural network cannot be exploited on the graph domain. The Deep Graph Convolutional Neural Network (DGCNN) (Sortpool), a global pooling architecture, solves this problem and learns from the graph's global topology. This architecture has three sequential stages:

1. The first stage consists of the Graph Convolution Layers, which abstract local substructure features and explicate ordering among the nodes of the graph. The graph convolution layer aggregates node features to draw information about local substructures. GCN equation can be described as follows:

$$H^{l+1} = \sigma(\hat{D}^{-1}\hat{A}H^{l}W^{l}) \tag{5.1}$$

- Here, $\hat{A} = A + I_N$ is the adjacency matrix of the graph after adding a self loop to each node of G. $\hat{D} \in \mathbb{R}^{N \times N}$ is the diagonal degree matrix. $W^{l} \in \mathbb{R}^{K \times K}$ is a trainable matrix, where K is the feature dimension of the hidden layers of GCN (except $W^0 \in \mathbb{R}^{D \times K}$) and $H^0 = X$. $H^{l} \in \mathcal{R}^{N \times K}$ is the output at lth graph convolution layer.
- Multiple layers are stacked, and the outputs of all L layers ($l = 1, \ldots, L$) are concatenated horizontally, which forms the output of this stage, denoted by $H^{1:L} = [H^1, H^2, \ldots, H^L]$. According to the authors, the graph convolution layers are analogous to two efficient kernels, Weisfeiler-Lehman subtree and propagation kernels, which explains its performance on classification. Further, the last layer output, H^L, can be exploited to order the nodes of a graph according to the structural roles of nodes.

2. The next stage is the SortPooling layer.

- This layer sorts the nodes' feature descriptors according to the node's structural roles.

  - The final colors in Weisfeiler-Lehman algorithm (WL) for isomorphism test can arrange the nodes in a consistent order based on their structural roles. That is, if two nodes from different graphs have similar structural roles, then those nodes are assigned the same relative position in their corresponding graphs. As a result, neural networks can read nodes in a meaningful, consistent order.
  - According to DGCNN, the graph convolution layers' outputs ($H^l, \forall l = 1, \ldots, L$) are the continuous WL colors; thus DGCNN algorithm uses these outputs to order the nodes of a graph. The input to the SortPooling layer is $H^{1:L} \in \mathcal{R}^{N \times KL}$, where each row represents a node, and each column represents a feature channel.
  - As the last layer output, H^L, is regarded as the most refined continuous WL colors; thus, the nodes are first sorted in descending order according to the values in the last channel of H^L. If two nodes have the same values in this channel, then the nodes are distinguished based on their values in second to last channel, and this is continued till the nodes have the same values. This arranges the nodes of a graph in a consistent order; hence, traditional neural networks can be applied to the sorted output.

- The next role of the SortPooling layer is to unify the sizes of the outputs. More precisely, the aim is to make the first dimension equal to M for all the graphs, making all the graphs of the same size, i.e., M nodes. This is done either by adding zero rows if $N < M$ or if $N > M$, then deleting $N - M$ rows from the last. Thus, the output of the SortPooling layer is a tensor $H^s \in \mathcal{R}^{M \times KL}$ where each row represents a node.

3. Third and the last stage is to train CNN and fully connected layers followed by a softmax layer. Convolution layers learn local patterns of the sequence, and the rest helps generate predictions for graph classification.

There are two significant advantages of DGCNN:

- First, the SortPooling layer of DGCNN behaves as a bridge between the first and last layers and can remember the sorted input. Hence, the loss gradients can flow back to the initial layers, and consequently, the parameters of these initial layers are also trained.
- Moreover, the complete DGCNN makes possible the use of traditional neural networks on graphs (Fig. 5.1).

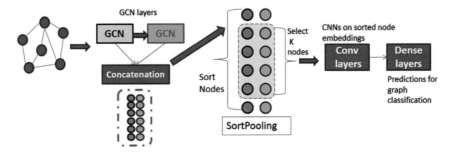

Fig. 5.1 An overview of SortPool architecture. First, two GCN layers are used to abstract local substructures features. Features of two GCN layers are concatenated and fed to the SortPooling layer, which sorts the nodes and selects K nodes. The output of this layer is passed through CNN and dense layers to perform graph classification

5.2 DIFFPOOL

Many real-world graphs have an inherent hierarchical graph structure. For example, in molecular graphs, both the local molecular and the coarse-grained structures are crucial to investigate the molecular graph structure. This hierarchical structure is important when the aim is to label the entire graph (Fig. 5.2).

A differentiable graph pooling (DIFFPOOL) module was proposed, which can maintain the hierarchical structure and can learn graph embedding for graph classification tasks in an end-to-end fashion.

- They generalize the pooling operation in CNN to the graph domain. Graphs, unlike images, have some challenges which make defining a graph pooling operator more demanding.

  - First, the node does not follow the spatial locality definition, which means the nodes in a patch cannot be pooled together. Moreover, the graphs in a graph collection dataset can have a different number of nodes and edges.

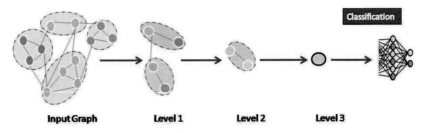

Fig. 5.2 An overview of DIFFPOOL architecture. At each layer, the input graph is coarsened using a GNN layer which pools the nodes. This process is continued until the graph is left with only one node, and this representation is used as input to the classifier for graph classification

- DIFFPOOL, an end-to-end differentiable strategy, overcomes these challenges and hierarchically stacks multiple GNNs. That is, it is a model to cluster the nodes to build a hierarchical multi-layered architecture and can be adapted to graphs of different sizes and with different structures.
- Each layer of deep GNN learns node embeddings, and based on these embeddings, determines the assignment of the current layer nodes to a set of clusters (differentiable soft assignment). Nodes at the lth layer of the DIFFPOOL are the same as the clusters generated at the $l-1$th layer.

Suppose the input graph is denoted by $G = (V, E)$ with a set of N nodes V and a set of edges E and is described by an adjacency matrix $A \in R^{N \times N}$ and node features matrix $X \in R^{N \times F}$ where F is the feature dimension.

Each successive GNN layer of DIFFPOOL learns a coarser version of the input graph. More precisely, given the output of GNN, $Z = GNN(A, X)$, and matrix A, DIFFPOOL determines how to learn a coarser version with less number of nodes defined by the weighted adjacency A' and the node embedding Z' matrices which are further fed to the next GNN layer. This process is continued for L times generating L layers of GNN, each one hierarchically pooling the nodes of the graph.

Suppose we denote the adjacency matrix and the feature matrix at layer l by $A_l \in \mathcal{R}^{N_l \times N_l}$ and $X_l \in \mathcal{R}^{N_l \times F}$ where N_l is the number of nodes in graph at layer l. We now discuss the lth layer of DIFFPOOL.

- The lth GNN embedding layer (between level l and $l+1$) is defined as:

$$Z_l = \text{GNN}_{l,embed}(A_l, X_l) \tag{5.2}$$

Here, $Z_l \in \mathbb{R}^{N_l \times F}$ is the embedding matrix of the nodes in lth level graph.
- Similarly, the assignment matrix is learned using the GNN pooling layer. This layer aims to assign the nodes from the previous layer to the clusters where each cluster is considered as a node in the next level graph. The output feature dimension of this layer is different from that of the embedding layer and is equal to the number of the nodes in the next level graph. Also, softmax is applied after the GNN pooling layer. The following equation describes this layer:

$$P_l = \text{softmax}(\text{GNN}_{l,pool}(A_l, X_l)) \tag{5.3}$$

Here, (i, j)th element of $P_l \in \mathbb{R}^{N_l \times N_{l+1}}$ is the probability of mapping node v_i^l to node (cluster) v_j^{l+1}. The softmax is used row-wise.
- Next, the adjacency and feature matrices are generated for the coarsened graph. The adjacency matrix A_{l+1} and node feature matrix X_{l+1} are constructed as:

$$A_{l+1} = P_l^T A_l P_l \in \mathbb{R}^{N_{l+1} \times N_{l+1}} \tag{5.4}$$

$$X_{l+1} = P_l^T Z_l \in \mathbb{R}^{N_{l+1} \times F} \tag{5.5}$$

- Matrix P_l contains information about how nodes in G_l are mapped to the nodes of G_{l+1}, and the adjacency matrix A_l contains information about the connection of nodes in G_l.
- These equations coarsen the graph. A_{l+1} represents the adjacency matrix for the next layer with N_{l+1} nodes. Each node in A_{l+1} is the cluster of the nodes in the lth level, and the ith row of matrix X_{l+1} represents cluster node i.

- This process is repeated until a graph with a single node is reached whose representation is used as the graph embedding.
- This representation is passed through a linear classifier to output the predictions for graph classification, and the parameters of the model are trained using stochastic gradient descent.

5.3 SAGPool

Self-Attention Graph Pooling (SAGPool) is a hierarchical pooling based model for GNNs. The model learns hierarchical representation for graph classification in an end to end-setting. The self-attention layer of SAGPool decides the nodes to be retained and the nodes to be dropped from the graph to output a smaller graph. Thus, each layer of SAGPool coarsens the graph by dropping some fraction of the nodes. SAGPool has four important layers:

1. **A Self-attention graph pooling layer**: This layer has two components.

   - A **Self-attention mechanism** to learn attention scores, which helps to decide the nodes to be kept for the next coarser version of the graph. The graph convolution layer is exploited to learn the self-attention scores $S^{N \times 1}$, which can be described as:

$$S = \sigma(\hat{D}^{-\frac{1}{2}} \hat{A} \hat{D}^{-\frac{1}{2}} X W) \tag{5.6}$$

$\hat{A} = A + I_N$ and $\hat{D} \in \mathbb{R}^{N \times N}$ is the degree matrix. $W \in \mathbb{R}^{D \times 1}$ is a convolution parameter and $X \in \mathbb{R}^{N \times D}$ is the node feature matrix with N nodes and D feature dimension. $S \in \mathbb{R}^{N \times 1}$ is the vector of attention scores where s_i is the attention of node i.

Next, SAGPool selects a fraction of nodes even when graphs in a graph classification dataset are of varying sizes. A hyperparameter, pooling ratio r, $0 < r \leq 1$, determines the portion of the nodes to keep, i.e., rN. The nodes are selected or discarded according to the learned self-attention scores, S. This can be described as follows:

$$topk = top - idx(S, rN)$$
$$S_{mask} = S_{topk} \tag{5.7}$$

Here $top - idx$ function returns the indices corresponding to the top $r N$ values, and $.idx$ denotes the index selection. S_{mask} represents the feature attention mask.

- The next component is the **graph pooling layer**. This layer coarsens the graph and generates the new adjacency matrix A' and the node features matrix X_{out}:

$$A' = A_{topk,topk}$$
$$X' = X_{topk,:}$$
$$X_{out} = X' \odot S_{mask} \tag{5.8}$$

Here, $X_{idx,:}$ is the row-wise indexing. $A_{mask,mask}$ is the row and column wise selections. $\odot$ is the element-wise dot product.

2. SAGPool also has a convolution layer to learn node embeddings, which again is a graph convolution layer as follows:

$$H^{l+1} = \sigma(\hat{D}^{-1}\hat{A}H^{l}W^{l}) \tag{5.9}$$

Here, $W^{l} \in \mathbb{R}^{D \times F'}$ is a trainable matrix, where D is the input feature dimension and F' is the output feature dimension. $H^{k} \in \mathcal{R}^{N \times F}$ is the node representation at kth graph convolution layer. σ is the ReLU activation function.

3. Finally, there is a READOUT layer. This layer aggregates the node features to generate a fixed size graph representation as below:

$$H_G = \frac{1}{N} \sum_{i=1}^{N} x_i || max_i^N x_i \tag{5.10}$$

Here, N is the number of nodes, $||$ is the concatenation and x_i is the ith node feature vector. H_G represents the summary output of the READOUT layer (Fig. 5.3).

SAGPool has two variants; the global and the hierarchical pooling architectures, which combine these four layers as follows:

- In global pooling, multiple graph convolution layers are stacked. All the outputs are concatenated, and a graph pooling layer is used to pool the nodes, followed by a READOUT layer to aggregate all the node embeddings.
- The hierarchical architecture contains multiple blocks where each block has a graph convolution and a pooling layer. Each block is followed by a READOUT layer to form a summarized output of the corresponding block. Lastly, the outputs of all the READOUT layers are summed together to generate a single vector, which is then passed through the classifier for graph classification.

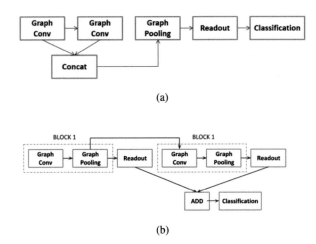

(a)

(b)

Fig. 5.3 A high-level illustration of global pooling and hierarchical pooling architecture. Graph Conv is graph convolution layer and the dense layer takes as input the graph representation and performs the graph classification

5.4   GIN

GNNs have achieved state-of-the-art performance on many node and graph classification tasks. They are revolutionizing the graph representation learning, but there is a limited theoretical understanding of representational power and shortcomings of GNNs. Existing GNNs are designed based on the empirical intuition or experimental trials. Graph Isomorphism Network (GIN) is a theoretical framework. Its discriminative or representational power is better than the existing GNNs and is equal to the WL test of graph isomorphism.

- GIN framework is motivated by the connection between GNNs and the WL isomorphism test, an efficient and powerful test for differentiating a wide range of graph structures.

  - The naive vertex alignment in the WL test is similar to the neighborhood aggregation in GNNs.
  - It is a two-step process: first, it aggregates the label information of the node and its neighbors, and second, it maps the new aggregated label to a new unique label.
  - Finally, if the labels of the nodes in the two graphs are not the same, they conclude that the graphs are non-isomorphic.
  - The injective behavior of its aggregation rule that maps different neighborhoods to different vectors makes WL so powerful test.

- Following the WL test, GIN proposes that if the aggregation step of GNN too is tremendously expressive and can model the injective function, then GNN can be as powerful as the WL test.

  - Precisely, the representational power of a GNN is proportional to the discriminative power of the aggregation function of the GNN. If the set of node features is represented as a multiset (a multiset can have repeating elements), then a discriminative GNN must map or aggregate different multisets to different vectors.
  - A powerful GNN should map two nodes with identical neighborhood structures having identical features to the same representations and should not map two different multisets of features (i.e., different graph structures) to the same representation vectors, i.e., should have an injective aggregation rule.

- Moreover, if a GNN maps two different graphs to different representations, the WL test will also decide the graphs as non-isomorphic.
- Further, to summarize, a GNN must satisfy the following constraints to be equally discriminative as the WL test:

1. The neighborhood aggregation function (i.e., a multiset of features) should be injective.
2. The graph-level readout function, which also operates over the multiset of features of the vertices, should also be injective.

Graph Isomorphism Networks (GIN):

- The proposed algorithm, Graph Isomorphism Network (GIN), satisfies the conditions above and has the maximum power of discriminating different graph structures.
- GIN proposes to use the MLP (multi layer perceptron) and the sum of the representations to map a multiset of node representations to new representations of nodes. GIN update rule is as follows:

$$z_v^{l+1} = MLP^l\left((1 + \epsilon^l)z_v^l + \sum_{u \in \mathcal{N}(v)} z_u^l\right) \tag{5.11}$$

  - Here, $z_v^{l+1} \in \mathbb{R}^K$ is the representation of the node v in $l + 1$th layer of GIN and K is the dimension of the representation. $\mathcal{N}(v)$ is the neighbors of the node v; ϵ is a trainable parameter which learns the significance of the node's representation relative to the aggregated features of its neighbors.
  - The Universal Approximation Theorem states that any function together with the injective function, can be approximated by the MLP. Thus, when the sum of node representations and the multi layer perceptron are injective, GIN is as powerful as the WL test of isomorphism.

- For graph classification, a READOUT function is further proposed, which operates over these learned representations of individual nodes and generates a single embedding representing the graph.

- At the later GIN layers, the node representation becomes more global, whereas sometimes previous layers' outputs may generalize better.
- Thus, to learn the graph embedding, GIN uses information from all the layers as follows:

$$Z^G = CONCAT\left(READOUT\left(z_i^l|i \in V\right)|l = 0, \ldots, L\right) \qquad (5.12)$$

Here, $V(G)$ is the set of vertices of graph G, L is the total number of layers in GIN.

5.5 Graph U-Nets

Graph U-Nets are encoder-decoder architectures based on the proposed graph pooling and graph unpooling layers. This model can be used for both node and graph classification tasks .

Unlike images, nodes of a graph are unordered and have no spatial locality because of which the existing pooling operations from the image domain cannot work on the graphs. Graph U-Nets provide a solution for this and propose gpool (graph pooling) and gUnpool (graph unpooling) layer (defined below).

gpool Layer: Pooling layers in CNN help in shrinking the size of the feature maps and increase the receptive fields. But because nodes have no spatial locality, these operations cannot be used on the graph domain.

Let us denote the adjacency matrix of a graph G as $A \in R^{N \times N}$ and the feature matrix as $X \in R^{N \times F}$ with N number of nodes and F feature dimension.

- The gpool layer introduces a trainable projection vector (p) that projects features of nodes onto 1-dimension. Using these projection scores, it performs k-max pooling and selects nodes with top k projection scores.
- The scalar score for each node v is computed as $s_v = \frac{x_v p}{||p||}$. This score describes the information node v can preserve when projected on p. Hence, by selecting the nodes with k largest scores, the gpool operation can retain the maximum information from the current graph.
- The lth propagation rule of gpool layer can be described as below:

$$S = \frac{X^l p^l}{||p^l||}$$
$$idx = rank(S, k)$$
$$\tilde{y} = sigmoid(S(idx))$$
$$X^l = X^l(idx, :) \odot (\tilde{y}1_F^T), \quad X^l \in R^{k \times F}$$
$$A^{l+1} = A^l(idx, idx), \quad A^{l+1} \in R^{k \times k} \qquad (5.13)$$

- Here k denotes the number of nodes in the smaller new graph. S is the vector where each entry is the projection score for a node on the vector p.

- In the next equation, the *rank* function ranks these scores in S and outputs the node's indices with k-largest scores.
- $S(idx)$ selects the scores from S at idx indices. 1_F is the unit vector of size F. $\odot$ denotes the element-wise multiplication. $X^l(idx, :)$ and $A^l(idx, idx)$ are row and column selection steps using indices set idx to generate the smaller graph's feature and adjacency matrices.
- Next, the gate operation is used, which brings each of the selected values between 0 and 1 by applying the sigmoid activation function. The vector of these scalars is denoted as $\tilde{y}$ in the above equations.
- Finally, $\odot$ operator is used to element-wise multiply the $X^l(idx, :)$ and $\tilde{y} 1_F^T$ that forms new feature matrix with information being controlled by the sigmoid gate.

gUnpool LAYER:

- An upsampling operation motivated by the up-sampling layers in encoder-decoder architectures, used in the image domain, is also proposed. The encoders shrink the sizes of the feature maps while the feature maps are up-sampled to the original size in decoders.
- gUnpool restores the original graph structure by performing the inverse operation of the graph pooling (gpool) layer.
- For this, the indices of the selected nodes are saved in the corresponding gpool layer. And the gUnpool layer uses these node indices to set the nodes into their actual place in the graph as follows:

$$X^{l+1} = distribute(0_{N \times F}, X^l, idx) \tag{5.14}$$

- Here $idx \in R^k$ denotes the index set from the corresponding gpool layer which reduces the number of nodes from N to k, $0_{N \times F}$ denotes a zero matrix of size $N \times F$ and $X^l \in R^{k \times F}$ describes the features of the nodes in the current graph.
- The function $distribute(.)$ places the matrix X^l in the empty node features matrix $(0_{N \times F})$ corresponding to the new graph at the indices stored in idx set. Thus, the rows of X^{l+1} corresponding to the indices in idx are the same as the rows of matrix X^l, and the remaining rows are zero.

The Complete Architecture of Graph U-Nets (g-U-Nets): The node classification problem is similar to the pixel-wise prediction tasks as both tasks try to predict for each input unit, such as pixel or node.

- In g-U-Net, a GCN embedding layer is first used to generate node embeddings in a lower dimension.
- Following the embedding layer, g-U-Nets stack multiple encoding blocks. This part is the encoder part of the g-U-Nets models.

  - An encoding block consists of one gpool layer and one GCN layer.
  - The graph pooling layer (gpool) decreases the graph size and captures higher-order features.

- The GCN layer aggregates features from each node's first-order neighbors and encodes the graph's topological information.

• The third part is the decoder part, which consists of several decoding blocks.

- The number of decoding blocks is the same as the number of encoding blocks in the encoder part, and each decoding block has a gUnpool and a GCN layer.
- The gUnpool layer, as discussed earlier, restores the original graph structure by performing the inverse operation of the graph pooling (gpool) layer, and the GCN layer performs the same operation as in the encoder part.

• g-U-Nets also consists of the skip connections, which are the addition or the concatenation of the feature maps, from the encoding block in the encoder part to the corresponding decoding block in the decoder part. These connections help in transferring the spatial knowledge from the encoding block to the decoding block.

• Lastly, a GCN layer is added for the prediction task, which is followed by the softmax layer.

As some fraction of nodes is selected, the rest of the nodes and the associated edges are thrown away, making some of the retained nodes isolated.

Thus to improve the connectivity among the nodes, it is further proposed to use the higher powers q, G^q. Graph-U-Nets set $q = 2$ i.e., $A^{l+1} = (A^l)^2(idx, idx)$, the 2nd graph power. They also change the GCN propagation layer by assigning more weights to the node's features than the node's neighboring nodes while computing the feature aggregation. The node's feature vector is more significant for making a prediction at the node. And thus, instead of adding self-loops, they calculate the adjacency matrix as $\hat{A} = A + 2I$ before normalization, giving higher weights to the node's features.

5.6 Experimental Evaluation

As stated earlier, these graph representations are used for various downstream tasks such as graph classification, graph clustering, visualization, etc. Any traditional ML algorithm can be used for these tasks by exploiting these learned vector embeddings as the input features. Further, the performance on these tasks can be used to assess the quality of the graph embeddings.

In this section, we assess the embeddings on graph classification and visualization tasks. We evaluate the performance of various algorithms on the publicly available benchmark datasets such as MUTAG, PTC, NCI, etc.

5.6.1 Graph Classification

Experimental Set-Up: For graph classification task, we use datasets of the bioinformatics domain and social network domain. MUTAG, PTC, PROTEINS, NCI1, and

NCI09 are bioinformatics graph datasets, and IMDB-BINARY and IMDB-MULTI are social network datasets. The details of these datasets are discussed in Sect. 2.5.

We compare the performance of all the algorithms discussed in this chapter, along with graph kernels. We perform 10-fold cross validation to compute the graph classification accuracy and record the standard deviation and averaged accuracy across ten folds. For unsupervised algorithm, node2vec, we take the average of all the node embeddings as the graph representation.

Performance Evaluation: Tables 5.1 and 5.2 show the graph classification performance (accuracy and standard deviation) of graph kernels and all the algorithms discussed in this chapter on different real-world graph datasets.

- We can observe that GIN achieves state-of-the-art performance on MUTAG and PTC datasets.
- The best performing algorithm among all the selected baselines on PROTEINS dataset is gpool.
- While on NCI datasets (NCI1 and NCI109), WL graph kernel is outperforming all the other algorithms (including the GNN based approaches and with a good margin (>1%)).
- On social network datasets (IMDB-BINARY and IMDB-MULTI), GIN is able to achieve state-of-the-art performance with a good performance gap from all the other selected algorithms (Table 5.2).

5.6.2 Visualization

In this section, we choose the graph visualizations as another downstream task.

Table 5.1 Classification accuracy (%) of different algorithms on Bioinformatic datasets. NA denotes the case when the result of a baseline algorithm could not be found on that particular dataset from the existing literature

Algorithms	MUTAG	PTC	PROTEINS	NCI1	NCI109
GK	80.99±1.7	54.74±0.2	70.84±0.2	61.88±0.5	61.95±0.7
RW	78.17±2.1	55.23±0.4	58.73±0.2	NA	NA
PK	75±2.7	59.0±2.5	72.89±1.0	83.19±0.4	NA
WL	83.97±1.9	58.29±3.2	74.23±0.3	**84.04±0.8**	**85.32±0.4**
node2vec	73.63±9.7	58.50±7.3	56.63±2.8	55.39±2.3	54.12±2.0
DGCNN	85.20±1.7	58.40±1.7	74.84±0.7	75.51±1.1	NA
DIFFPOOL	84.61	63.5	76.25	NA	NA
SAGPool	80.9	62.3	71.4	74.2	74.5
gpool	79.84	NA	**77.5**	NA	NA
GIN	**89.4±5.6**	**64.6±7.0**	76.2±2.8	82.7±1.7	NA

Table 5.2 Classification accuracy (%) of different algorithms on Social Network datasets. NA denotes the case when the result of a baseline algorithm could not be found on that particular dataset from the existing literature

Algorithms	IMDB-B	IMDB-M
AWE-DD	73.66±4.9	51.00±3.2
AWE-FB	73.20±2.9	51.23±3.8
DGCNN	71.03±0.9	48.21±1.2
DIFFPOOL	73.3	50.0
SAGPool	72.80	50.80
gpool	73.2	49.8
GIN	**75.1±5.1**	**52.3±2.8**

- The graph representations are learned using any graph representation learning algorithm and fed to t-SNE, which converts them into 2D planes.
- Distinct colors denote different classes (labels) of the graphs. The performance is better if distinct colors form separate clusters in the plot.
- We select a subset of algorithms, including pooling based and GNN based algorithms (DIFFPOOL, SAGPool, and GIN) and a subset of datasets (MUTAG, PTC, and IMDB-BINARY).
- Figure 5.4 shows the visualization of the graphs of the MUTAG dataset. Figure 5.5 shows the visualization of the graphs of the PTC dataset, and Fig. 5.6 depicts the visualization of the graphs of the IMDB-BINARY dataset.

5.6.2.1   Performance Analysis

- On MUTAG and PTC, GIN (refer to Figs. 5.4c and 5.5c) achieves better performance as compared to DIFFPOOL (Figs. 5.4a and 5.5a), and SAGPool (Figs. 5.4b and 5.5b).
- Also, on the MUTAG dataset, the performance in Fig. 5.4b is slightly better than the performance of DIFFPOOL in Fig. 5.4a. A similar observation is also prominent in Fig. 5.5 on the PTC dataset.
- While on the social network (IMDB-BINARY) dataset, the best performance is observed in Fig. 5.6b, which uses SAGPool algorithm. Further, the performance of GIN in Fig. 5.6c is better than the performance of DIFFPOOL in Fig. 5.6a.

(a) DIFFPOOL   (b) SAGPOOL   (c) GIN

Fig. 5.4 t-SNE visualization of the graphs of MUTAG dataset (distinct colors depict different graph labels) by the embeddings produced by different algorithms

(a) DIFFPOOL   (b) SAGPOOL   (c) GIN

Fig. 5.5 t-SNE visualization of the graphs of PTC dataset (distinct colors depict different graph labels) by the embeddings produced by different algorithms

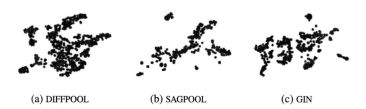

(a) DIFFPOOL   (b) SAGPOOL   (c) GIN

Fig. 5.6 t-SNE visualization of the graphs of IMDB-BINARY dataset (distinct colors depict different graph labels) by the embeddings produced by different algorithms

Bibliography

1. Hamilton WL, Ying R, Leskovec J (2017) Representation learning on graphs: methods and applications. arXiv preprint arXiv:1709.05584
2. Zhang D, Yin J, Zhu X, Zhang C (2018) Network representation learning: a survey. IEEE Trans Big Data
3. Zhang M, Cui Z, Neumann M, Chen Y (2018) An end-to-end deep learning architecture for graph classification. In: Thirty-second AAAI conference on artificial intelligence
4. Lee J, Lee I, Kang J (2019) Self-attention graph pooling. arXiv preprint arXiv:1904.08082
5. Ying Z, You J, Morris C, Ren X, Hamilton W, Leskovec J (2018) Hierarchical graph representation learning with differentiable pooling. In: Advances in neural information processing systems, pp 4800–4810
6. Gao H, Ji S (2019) Graph u-nets. arXiv preprint arXiv:1905.05178
7. Xu K, Hu W, Leskovec J, Jegelka S (2018) How powerful are graph neural networks?. arXiv preprint arXiv:1810.00826

8. Sun FY, Hoffmann J, Verma V, Tang J (2019) Infograph: unsupervised and semi-supervised graph-level representation learning via mutual information maximization. arXiv preprint arXiv:1908.01000
9. Maron H, Ben-Hamu H, Serviansky H, Lipman Y (2019) Provably powerful graph networks. In: Advances in neural information processing systems, pp 2156–2167

Chapter 6
Conclusions

In this book we have examined *social and information networks*, and their analysis. Specifically, we considered the following aspects.

1. A fundamental problem in *data analysis* is representation. So, *representation learning* is the most important step in dealing with almost any large-scale practical problem.
2. In this book we have examined in detail different schemes for *network representation learning (NRL)*.
3. There was more emphasis on *social and information networks* in the book. However, the schemes discussed are generic and can be applied to any other complex network.
4. It is important to note that data in the form of networks is either explicit or implicit where the networks are typically represented as graphs.
5. The importance of *networks* in dealing with any application need not be over emphasized. They are so important that the schemes considered in the book are useful in both implicit and explicit cases.
6. The basic problem examined in detail in the book is embedding network entities. Both node and graph embedding schemes are examined in detail. Further, state-of-the-art embedding schemes are compared using several benchmark datasets.
7. The background required in terms of graphs, adjacency matrices, matrix factorization, random walks, representing words as vectors, neural networks, and deep learning schemes are discussed in detail in Chaps. 2 and 3.
8. Evaluation of various embedding schemes is typically done with the help of downstream ML tasks including classification, community detection, link prediction and visualization. We have explained these ML tasks in Chap. 2.

© The Author(s), under exclusive license to Springer Nature Singapore Pte Ltd. 2021   105
M. Aggarwal and M. N. Murty, *Machine Learning in Social Networks*,
SpringerBriefs in Computational Intelligence,
https://doi.org/10.1007/978-981-33-4022-0_6

9. Different schemes for embedding nodes in a network are examined in Chap. 4. In Chap. 5, various schemes for embedding an entire graph are considered.
10. A brief summary of the importance of networks and their representations is done in the current chapter with a view that networks will play an important role, in every practical application, in the near future.

Glossary

G	Graph representing a network
V	Set of vertices or nodes in a graph
E	Set of edges in a graph
N	Number of nodes in a graph
A	Adjacency matrix of the graph
v_i	ith node of the graph
$e - ij$	edge between nodes v_i and v_j
x_i	attribute associated with node v_i
X	Set of node attributes
l_i	Label associated with node v_i
L	Set of node labels
$\mathcal{G} = \{G_1, G_2, \ldots, G_M\}$	Set of graphs
L_g	Set of graph labels
$x - j^i \in \Re^D$	Attribute vector for the jth node in the ith graph

© The Author(s), under exclusive license to Springer Nature Singapore Pte Ltd. 2021   107
M. Aggarwal and M. N. Murty, *Machine Learning in Social Networks*,
SpringerBriefs in Computational Intelligence,
https://doi.org/10.1007/978-981-33-4022-0

Index

© The Author(s), under exclusive license to Springer Nature Singapore Pte Ltd. 2021   109
M. Aggarwal and M. N. Murty, *Machine Learning in Social Networks*,
SpringerBriefs in Computational Intelligence,
https://doi.org/10.1007/978-981-33-4022-0

Printed in the United States
By Bookmasters